Dr. Math® Explains

ALGEBRA

Dr. Math® Explains

ALGEBRA

• • • • • • • • • •

Learning Algebra Is Easy! Just Ask Dr. Math!

THE MATH FORUM

Cartoons by Jessica Wolk-Stanley

John Wiley & Sons, Inc.

This book is printed on acid-free paper. ♾

Published by John Wiley & Sons, Inc., Hoboken, New Jersey
Published simultaneously in Canada

Design and production by Navta Associates, Inc.

ISBN 0-471-22555-X

Printed in the United States of America

10 9 8 7 6 5 4 3 2 1

Contents

Factoring 93

Quadratic Equations 123

Acknowledgments........................

Suzanne Alejandre and Melissa Running created this book based on the work of the Math Doctors, with lots of help from Math Forum employees, past and present:

Ian Underwood, Attending Physician

Sarah Seastone, Editor and Archivist

Tom Epp, Archivist

Lynne Steuerle and Frank Wattenberg, Contributors to the original plans

Jay Scott, Math Consultant

Kristina Lasher, Associate Director of Programs

Stephen Weimar, Director of the Math Forum

We are indebted to Jerry Lyons for his valuable advice and encouragement. Our editors at Wiley, Kate Bradford and Kimberly Monroe-Hill, have been of great assistance.

Our heartfelt thanks goes out to the hundreds of Math Doctors who've given so generously of their time and talents over the years, and without whom no one could Ask Dr. Math. We'd especially like to thank those doctors whose work is the basis of this book:

Anthony Hugh Back, Pat Ballew, Toby Bartels, Ezra Brown, Michael F. Collins, Bob Davies, Tom Davis, Sonya Del Tredici, Bonnie Devine, Patrick Donahue, Concetta Duval, C. Kenneth Fan, Scott Fellows, Sydney Foster, Sarah Seastone Fought, Loni Ghiorso, James Gill, Margaret Glendis, Aaron Hoffman, Bill Homann, Maureen Giglio Honeychuck, Jerry Jeremiah, Floor van Lamoen, Ethan Magness, Jerry Mathews, Josh Mitteldorf, Paul Narula, Elise Fought Oppenheimer, Andrew Parker, Dave Peterson, Richard Peterson,

Otavia Propper, Beth Schaubroeck, Jodi Schneider, Santu de Silva, Gary Simon, Steven Sinnott, Mark Snyder, Alicia Stevens, Gary Stoudt, Betsy Teeple, Sandra Trewartha, Ian Underwood, Jody Underwood, Joe Wallace, Peter Wang, Robert L. Ward, Elizabeth Weber, Lisa Widman, John Wilkinson, and Ken Williams.

Drexel University graciously hosts and supports The Math Forum, reflecting Drexel's role as a leader in the application of technology to undergraduate and graduate education.

Introduction

Students have been writing to Dr. Math® for years asking questions about how to figure out math problems using algebra, and Dr. Math has been helping them by replying with clear explanations and helpful hints. When you have spent time working through this book and our companion book, *Dr. Math Gets You Ready for Algebra,* you will have the help you need to succeed in your algebra course.

In algebra, there is a jump from the concrete world of numbers and real objects you recognize, to the abstract world of letters and symbols. To quote Dr. Math, "Algebra is the class where you learn how to work with unknown quantities." Generally in Algebra I, a student learns how to solve linear equations (equations with just an x in them), graph straight lines, set up some word problems, factor, and perform operations (such as adding, subtracting, multiplying, and dividing) on rational expressions.

Here are some general pointers from our experienced Math Doctors:

1. If you have an equation and you do *exactly* the same thing to each side of the equation, what you wind up with will be a true equation. Here is a simple example: Given $2x + 6 = 32$, if you subtract 6 from both sides, you get $2x = 26$. If you then divide both sides by 2, you get $x = 13$. Since the first equation is given to be true, the equation $x = 13$ is also true.

2. Write everything down. Don't do stuff in your head. You may be tempted for the previous example to say, "Okay, I'll subtract 6 and divide by 2 and get, ummm . . . x = 14." Paper is cheap. Write it down. Go step-by-step. Don't cut corners. There are two reasons to take this advice. It is good advice because it helps you solve problems when you can see everything on

paper that you're working with. Also, teachers really love to see work. Really!

3. Always check your answer. If you get the answer 14, go back and check to see if it really is a solution. Do the math! Find $2 \cdot 14 + 6$ and see if it really is 32. If it is not, you have messed something up and it's your responsibility to find out what.

4. Check each step. It's really easy even for a Ph.D. in math to multiply $6 \cdot 7$ and get 56 if he or she is not being careful. These are just details, but a lot of math *is* details. There are some important concepts, but there are lots of details. In order to learn the concepts, you must sweat the details.

5. Think first about what the problem means. If you are asked to answer, "What is the square root of $2357 \cdot 2357$," and you know what a square root is, then you should be able to do that in your head. Problems dealing with exponents are prime opportunities to use this hint. For example, if you are asked to simplify "4 to the 1.5 power," you might think, "calculator time"; but that's *not* the best way. This problem is an opportunity to actually use one of the math concepts you learned. Since 1.5 is $\frac{3}{2}$, the problem is really "4 to the $\frac{3}{2}$ power." Now you *must* review the definition of fractional powers. Taking something to the $\frac{3}{2}$ power *means* to "cube it to get a result, then square root that result." No calculator needed yet, just words and what they mean. Now we know what the problem is really asking. Simplify: $\sqrt{4^3}$. What's 4 cubed? It's 64, right? What's the square root of 64? You probably know that, but I suppose you *could* use your calculator to get 8. A good rule of thumb for use of your calculator in algebra is to use it at most once per problem. Save that one time for when you really need it.

6. This was mentioned before, but it's important enough to get its own number. *Sweat the details.* If you make "silly" arithmetic mistakes in doing an exercise about a new concept you are learning, then you might think that you do not understand the new concept. Algebra has some interesting ideas and

techniques, so keep your arithmetic skills in order, and you'll see them clearly.

7. Reread this introduction at least once a week for the rest of the year. This *will* help you in algebra.

Linear Equations

Linear equations (equations that describe a line) are the simplest equations that mathematicians study. The work you've done with integers and coordinate graphing will help you with the topic of linear equations. If you have two points on a graph, you can connect them to form a line. When you see a line on a graph, you might notice some of the points it goes through. You could be precise in naming

those points by identifying both the x- and y-coordinates. You might also notice if the line is slanted to the left or to the right or if it is horizontal or vertical. Both the location of the points on the line and the slant of the line are important characteristics used to describe a line.

To describe a line algebraically, you write *an equation of the line*. When you speak of those equations in general, you are referring to *linear equations*. Once you're familiar with some examples, you can tell a lot about what the graph of an equation will look like just by looking at the equation.

> **Rx FOR HELP**
>
> **If you're not clear on integers or coordinate graphing, it's time to go to *Dr. Math Gets You Ready for Algebra*. Work through some of the questions and answers in that book before you get started here.**

Because linear equations are the most simple examples of the equations that mathematicians study—and because they are so useful for describing things that go on in the world—mathematicians have developed many tricks and shortcuts for dealing with them. In this chapter, we'll learn some of those tricks and shortcuts.

In this part, Dr. Math explains:

- linear expressions and equations
- slope, intercepts, and slope-intercept form
- graphing linear equations

1 Linear Expressions and Equations

Linear expressions and linear equations are related concepts. A linear expression is an expression with a variable in it; however, the variable is raised only to the first power. For example, $5a$ is a linear expression. Another example is $5a + 2$. If instead you had $5a + 2 = 12$, then you would have an example of a linear equation. What's important to remember is that an *equation* has an equal sign but an *expression* does not.

Linear Expressions

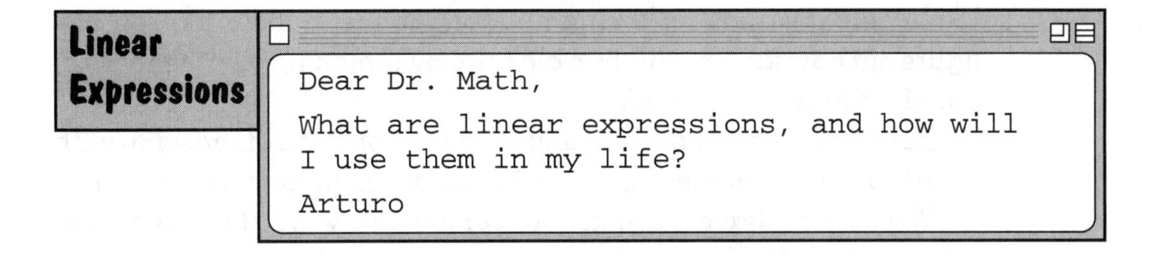
Dear Dr. Math,

What are linear expressions, and how will I use them in my life?

Arturo

Dear Arturo,

Thanks for your question. Let me start with the "what are they" part of your question.

I am going to explain mathematical expressions by comparing them to something you probably understand from studying English grammar in school. When we write, we use sentences to write a complete thought. In a sentence, there must be a noun and a verb, and often there are extras like descriptive words.

In mathematics, we also frequently write in sentences, but we use numbers and symbols to convey a thought. A complete mathematical sentence includes an equal sign or inequality sign ($<$ or $>$) and at least one term on either side. For instance: $5 + 3 = 8$ is a mathematical sentence, called an **equation,** while $9 < 100$ is also a mathematical sentence, called an inequality.

When we write, we may also use phrases, which are groups of words that are not complete sentences, like "in a nutshell" or "good sport." In mathematics, we use phrases too, but they're called **"expressions."** Expressions can be just one number or several numbers and some symbols; however, there is no equal sign or inequality sign between them. For example, 8 is an expression, and so are $\frac{4}{3}$ and $5 + 3$.

Some mathematical expressions include letters that stand for something else. These letters are called **variables** because they represent numbers that can vary. Expressions with variables are used every day in all sorts of situations. Here is an example: Let's say that you have a car that can travel 15 miles for every gallon of gas in the tank. You could represent the total number of miles you can drive based on how much gas you have in the tank using the expression

15g, where g stands for the number of gallons in the tank. You can figure out how far you will be able to go by replacing g with the number of gallons in your tank.

Now for the linear part. A linear expression is an expression with a variable in it, but there is a special condition involving exponents.

You may not have learned about exponents yet, so I'll give a brief explanation. We use exponents to symbolize many multiplication operations using the same number. For instance, perhaps you must multiply 3 by 3 by 3 by 3 by 3. We would write this as 3^5, which means the product of five 3's. We read the expression 3^5 as "3 raised to the fifth power." Any number raised to one is simply that number: $4^1 = 4$; $5a^1 = 5a$, and so on. We usually omit writing the "^{1}" when a number is raised to the first power.

A linear expression is an expression with a variable in it; but only when the variable is raised to the first power. For example, $5a$ is a linear expression, because it is understood that a is raised to the first power, but $9t^2 + 8$ is not a linear expression because t is raised to the second power.

There are many possible examples of linear expressions that you might use in your life. Another common one is computing how much money you might earn in a week working at a restaurant where you were paid by the hour. If your salary is $6 per hour, your total pay for a week could be expressed as $6h$, where h represents the number of hours you worked during that week. Can you think of more examples now?

— Dr. Math, The Math Forum

Equation, Function, or Formula?

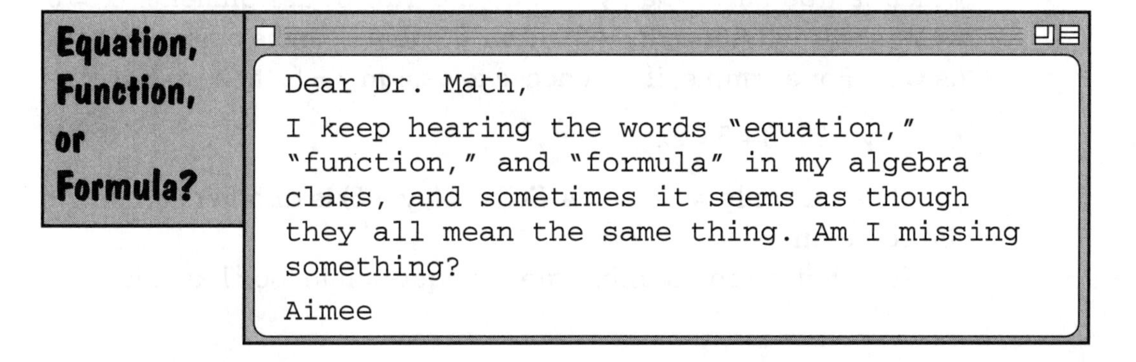

Dear Dr. Math,

I keep hearing the words "equation," "function," and "formula" in my algebra class, and sometimes it seems as though they all mean the same thing. Am I missing something?

Aimee

Dear Aimee,

Good question! It's not something that gets discussed as much as it should, so I'm glad you asked about it.

An *equation* is simply an assertion that two quantities are equal, for example,

$$3 + 5 = 8$$

When we include a variable in an equation, for example,

$$3 + x = 8$$

we're still asserting that two quantities are equal, but now we're doing something more. We're also asserting that there is some value (or set of values) that can be assigned to the variable to make the equation true. In the equation above, only the assignment $x = 5$ will make the equation true.

When we include a second variable, for example,

$$3 + y = 2x$$

we're asserting that there are *pairs* of assignments that make the equation true. That is, we can *choose* a value for one of the variables, and that choice will *determine* a value (or values) for the other variable.

For example, suppose we choose to assign $x = 5$ in the equation above. Then we have:

$$x = 5: \quad 3 + y = 2(5)$$

which is true only when $y = 7$. So the pair of assignments ($x = 5$, $y = 7$) is one solution to the equation. But there can be other solutions as well. For example, if we choose to assign $y = 5$, then we have:

$$y = 5: \quad 3 + 5 = 2x$$

which is true only when $x = 4$. So ($x = 4$, $y = 5$) is another solution to the equation.

Note that we can manipulate the equation to look like this:

$$y = 2x - 3$$

This has the same meaning, but it's somewhat easier to work with. Now if we choose a value for x, we can evaluate the resulting expression, and that gives us the corresponding value for y:

Choose	**Evaluate**	**Solution**
$x = 1$	$y = 2(1) - 3 = -1$	$(x = 1, y = -1)$
$x = 2$	$y = 2(2) - 3 = 1$	$(x = 2, y = 1)$
$x = 3$	$y = 2(3) - 3 = 3$	$(x = 3, y = 3)$

When we write the equation in this way, we say that y is a **function** of x. That is, it's a kind of rule for starting with one value (the "input") and finding a matching value (the "output"). Often, when we turn an equation into a function, we'll simply drop the output variable and use a slightly different notation. Instead of

$$y = 2x - 3$$

we'll write

$$f(x) = 2x - 3$$

But it means the same thing.

So we've talked about equations and about functions. What about formulas? A **formula** is simply an equation that is so useful that we want to share it with other people. It's often written in the form of a function, although it's not restricted to that use. For example, the formula for the volume of a cylinder is:

$$V = \pi r^2 h \quad V = \text{volume}, r = \text{radius}, h = \text{height}$$

Written this way, it tells us volume as a function of radius and height. But sometimes we already know the volume and the radius, and we need to find the height! We don't bother to make up a new formula, because we can just change this one around to find what we want. So a formula is often a function, but it doesn't have to be, and we don't have to use it that way. Mainly a formula is an equation that is useful enough to write down in a permanent location (like a book or a Web site), so that we can look it up instead of having to figure it out from scratch each time we want to use it.

—*Dr. Math, The Math Forum*

2 Slope, Intercepts, and Slope-Intercept Form

We talk about "ski slopes" and how steeply a roof "slopes" when we're referring to something that is not horizontal or vertical but at a slant. In math we use the term **slope** similarly. It is a number that tells how steeply a line slants as it goes up or down. Slope is important, because if you know the slope as well as the location of any point on a line, you have everything you need to know to find all of the other points on the line. So often, when you're given some information about a line, the first thing you want to do is figure out what the slope is.

When we talk about intercepts, it makes sense to to look at a graph. In the Cartesian system of graphing, we have the x-axis and the y-axis. When we graph a line, it may cross the x-axis or the y-axis, or both. Any point at which it crosses an axis is called an **intercept.** The point where it crosses the x-axis is known as the **x-intercept,** and the point will be in the form (x, 0). The point where it crosses the y-axis is known as the **y-intercept,** and the point will be in the form, $(0, y)$.

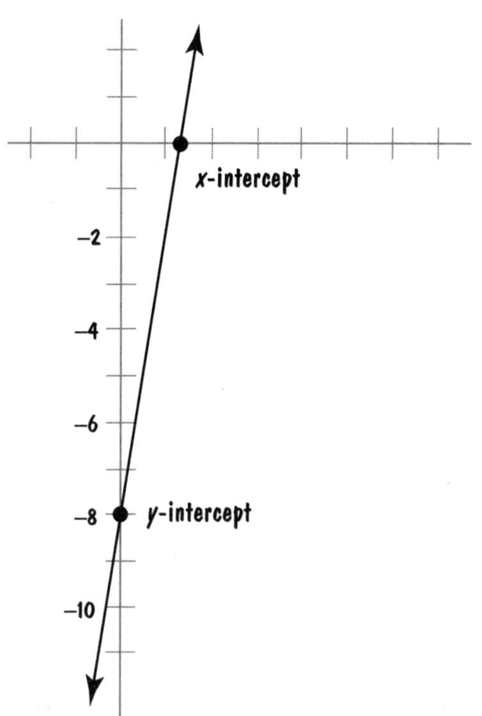

You can see how useful slope and intercepts are for describing or graphing a line. Because of this, one of the standard forms for linear equations shows slope and y-intercept clearly. It's called the slope-intercept form, and it's written:

$$y = mx + b$$

In this equation, the m stands for the slope of the line, and the b stands for its y-intercept. So this line has slope m and crosses the y-axis at $(0, b)$.

What Is Slope?

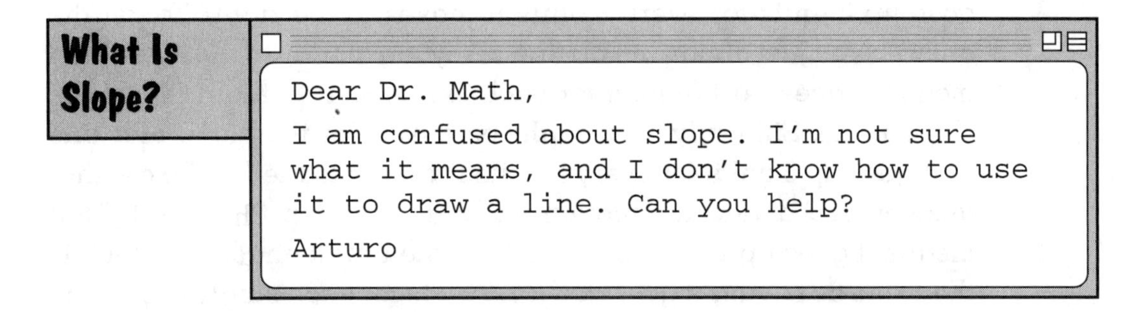

Dear Dr. Math,

I am confused about slope. I'm not sure what it means, and I don't know how to use it to draw a line. Can you help?

Arturo

Dear Arturo,

The slope is just a number that tells how steeply a line goes up or down. If the line is perfectly level (it doesn't go up or down at all), the slope is zero. If, as you go to the right, the line gets higher, we say it is sloping up, or has a positive slope. If it goes down as we go to the right, we call that a negative slope.

So that's how to think of slopes. Here's how to measure (or draw) them. Let's first look at specific examples.

Suppose the slope is 1. That means that if you go 1 unit to the right, the line goes up by 1 unit. The units can be whatever you choose, so if you go 1 foot to the right, the line goes up 1 foot. If you go 1 centimeter to the right, the line goes up 1 centimeter, and so on. If you draw this line, you'll find it goes up at a 45-degree angle.

If the slope is 3, the line goes up more steeply. If you go 1 unit to the right, the line goes up 3 units, where "unit" can be inch, foot, centimeter, or whatever.

If the slope is $\frac{1}{2}$, it means that if you go 1 unit to the right, the line goes up $\frac{1}{2}$ unit. If the slope is 1,000, the line is very steep—going 1 unit to the right, the line rises by 1,000 units, and so on.

If the slope is negative, it works the same way, except the line goes down to the right. A slope of -1 means that going 1 unit to the right, the line drops 1 unit. A slope of $-\frac{1}{3}$ means for every unit the line goes to the right, it drops by $\frac{1}{3}$ of a unit, and so on.

There's one nasty problem and that concerns lines that go straight up and down—you can't assign a sensible slope to them, because they never go to the right. We call these slopes **undefined.**

Often you'll get problems like this: What's the slope of a line that goes up 3 units for every 2 units it moves to the right? To get the answer, you just divide the motion up by the motion to the right. That means it goes up 1.5 units for each single unit to the right, so the slope is 1.5. Similarly, you might be asked for the slope of a line that goes up 3 units for every 2 units it moves to the *left*. To get that answer, you divide the motion up by the motion to the left. That means it goes up 1.5 units for each single unit to the left—which is the same as saying it goes *down* 1.5 units for each single unit to the *right*, so the slope is –1.5.

—Dr. Math, The Math Forum

Point-Slope Equations

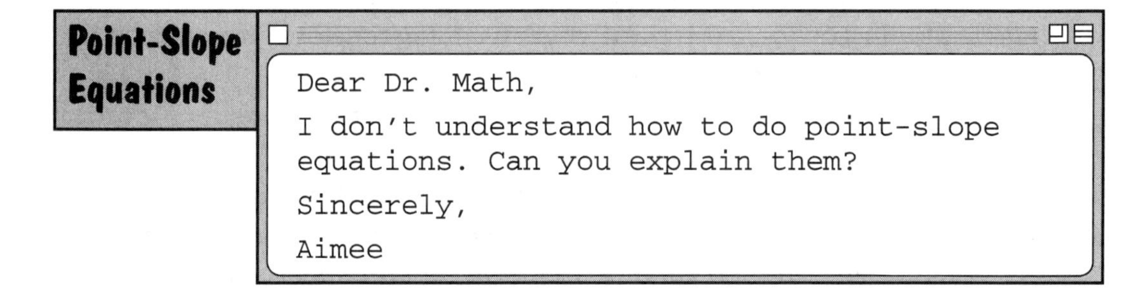

Dear Dr. Math,

I don't understand how to do point-slope equations. Can you explain them?

Sincerely,

Aimee

Dear Aimee,

The point-slope form of an equation is just one of many ways to write the equation of a line. It's handy to use when you know the slope of a line and one point on it.

A point-slope equation looks like this:

$$y - y_1 = m(x - x_1)$$

where m is the slope and x_1 and y_1 correspond to a point on the line.

In order to solve a problem (that is, write an equation of a line) using the point-slope equation, you need two things: a point on the line (x_1, y_1) and the slope of the line.

For example, to find the equation of a line with a slope of 2 and a point on the line (–1, 3), m would be equal to 2, x_1 would be –1, and y_1 would be 3.

Plugging them into the point-slope equation, you get:

$$y - 3 = 2(x - (-1))$$

Then solve for y to simplify the equation.

$$y - 3 = 2(x + 1)$$

$$y - 3 = 2x + 2$$

$$y = 2x + 5$$

Sometimes you will get a problem that says to write the equation of the line, but you are only given two points and not the slope. For example, find the equation of the line that contains the points (2, 1) and (0, –1). In order to use the point-slope form of the equation, you need to find the slope (m), which is the difference in the y-coordinates divided by the difference in the x-coordinates (i.e., "rise over run"— see Rx on the next page):

$$m = \frac{y_2 - y_1}{x_2 - x_1}$$

Putting in the numbers, you would get

$$m = \frac{-1 - 1}{0 - 2} = \frac{-2}{-2} = 1$$

Therefore the slope of the line is 1.

Notice that I subtracted the coordinates of (x_1, y_1) from the coordinates of (x_2, y_2). You might ask what would have happened if I had done it the other way and subtracted the coordinates of (x_2, y_2) from the coordinates of (x_1, y_1), giving

$$\frac{y_1 - y_2}{x_1 - x_2}$$

The result would be the same. Just be sure that you use the same order for both the numerator and the denominator.

RISE OVER RUN

"Rise over run" refers to slope, and it describes the up/down movement on a graph as compared to the right/left movement on a graph. You determine the up/down movement on a graph by comparing the *y*-coordinates of two points on the line. You determine the left/right movement on a graph by comparing the *x*-coordinates of two points on the line. This is also sometimes described as "change in *y* over change in *x*."

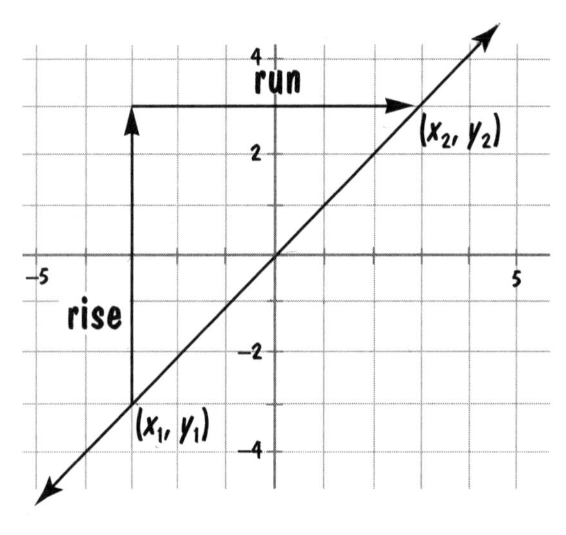

Using the slope and one of the points you were given (it doesn't matter which one), you can use the point-slope form (I'll use the first point given):

$$y - 1 = 1(x - 2)$$

Solve for y to simplify.

$$y - 1 = x - 2$$

$$y = x - 1$$

—*Dr. Math, The Math Forum*

When Is a Slope Zero or Undefined?

Dear Dr. Math,

How do I know when the slope of an equation is zero or undefined (no slope)? Thank you!

Sincerely,

Aimee

Dear Aimee,

Since you didn't say, I'll assume that by "equation" you mean the equation of a line—that is, a linear equation. Talking about slopes gets a great deal more complicated when the equation isn't linear.

Using the slope-intercept form, $y = mx + b$, if the slope of the line is zero, then this means that $m = 0$. But if m is zero, then mx is also zero, since anything times zero must be zero. So we are left with:

$$y = b$$

This is the only way to have a line with slope 0. If we graph this line, what do we get? Let's take a concrete example:

$$y = 1$$

This is a perfectly good line. Notice that x is missing. But we know that y is always equal to 1. So all the points on this line will have 1 as their y-coordinate. Since x is missing, it may be any value at all:

$$(1, 1), (2, 1), (3, 1), (4, 1), (1000, 1), (-2034, 1) \ldots$$

All these are on the line. What does this look like if we graph it? Try it, and you'll see that it is a horizontal line going through (0, 1). The x-value can be any number, but y must always be 1.

So what do we conclude? All horizontal lines have a slope of zero. And if any line has slope of zero, it must be a horizontal line.

This makes sense, given what we know about slope. A line may have a positive slope, negative slope, zero slope, or undefined slope.

The way to tell which slope a line has is to do this: Pretend you are standing on the line, on its graph, and you are going to walk from *left* to *right* along the line. If you are:

walking *up* hill: it's a positive slope

walking *down* hill: it's a negative slope

walking a *flat* line: it's a zero slope with no hill at all

Now, that leaves only the undefined slope. In the case of a line, this means that the line is *vertical*. Remember, a horizontal (flat) line has a slope of zero. A vertical line (one that forms a right angle or is perpendicular to a flat line) has an undefined slope. Think of it using the hill analogy again. If the hill is straight up, we couldn't walk up it. If and only if the line is perfectly vertical do we say it has an undefined slope. If it's almost vertical, but not quite, then it will have a very big (steep) slope but not be undefined.

Why is this so? I'll try to explain.

Imagine you have drawn a vertical line on your graph paper, through the point (1, 0).

To get to the next point on the line, which is (1, 1), what do you have to do? You have to go up 1 and over none. That is, the rise is 1 and the run is 0.

So you have the fraction $\frac{1}{0}$. Aha! Now do you see why a vertical line has an undefined slope? Because you can't divide by zero!

Usually in math, when something is "undefined" it means that somewhere, something is being divided by zero. And that's a no-no. It has no meaning. So we call it undefined.

—*Dr. Math, The Math Forum*

REMEMBERING *M* AND *B*

Robby Grant, a student, has suggested a way of remembering *m* for slope and *b* for *y*-intercept:

> **I think of *m* as standing for "move" and *b* for "begin." This relates to the way you graph linear equations by hand. You can use the *b* value to plot the "beginning" point (0, *b*). Then the *m* value instructs you where to "move" from point (0, *b*) to plot the next point, thus giving you the line for the equation.**

What Is a *Y*-Intercept?

Dear Dr. Math,

I'm having a problem with finding *y*-intercepts. How do you know what the *y*-intercept is for the following equations?

$2x + y = 3$ or $x - 4y + 8 = 0$

Arturo

Dear Arturo,

The *y*-intercept is the place where the graph hits the *y*-axis. The *y*-axis represents all the points on the plane where the x value is zero.

Putting this another way, to find a point given the x- and *y*-coordinates, you move left or right by the x amount, and then up or down by the *y* amount. To stay on the *y*-axis, you make no motion in the x direction; in other words, the x-coordinate is zero.

So the y-intercept of $2x + y = 3$, occurs when $x = 0$. This means that $2 \cdot 0 + y = 3$, or $y = 3$, so the y-intercept of this equation is 3. For $x - 4y + 8 = 0$, when you set $x = 0$, you get $4y + 8 = 0$, or $4y = 8$, or $y = 2$. That's all there is to it.

—Dr. Math, The Math Forum

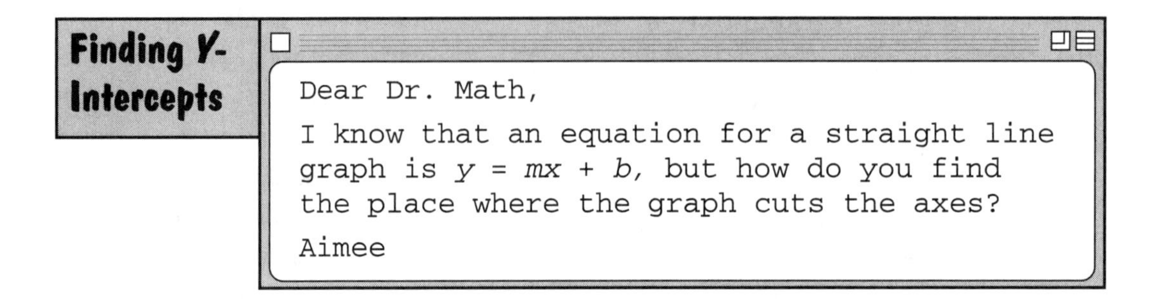

Finding Y-Intercepts

Dear Dr. Math,

I know that an equation for a straight line graph is $y = mx + b$, but how do you find the place where the graph cuts the axes?

Aimee

Dear Aimee,

If you have an equation in the form $y = mx + b$, the slope is m, and it intersects the y-axis at the point $(0, b)$ because the y-axis is where x is zero. To find b, you plug in zero for x and you get $y = m \cdot 0 + b = b$. To find out where the line hits the x-axis, you'd plug in zero for y and solve for x.

—Dr. Math, The Math Forum

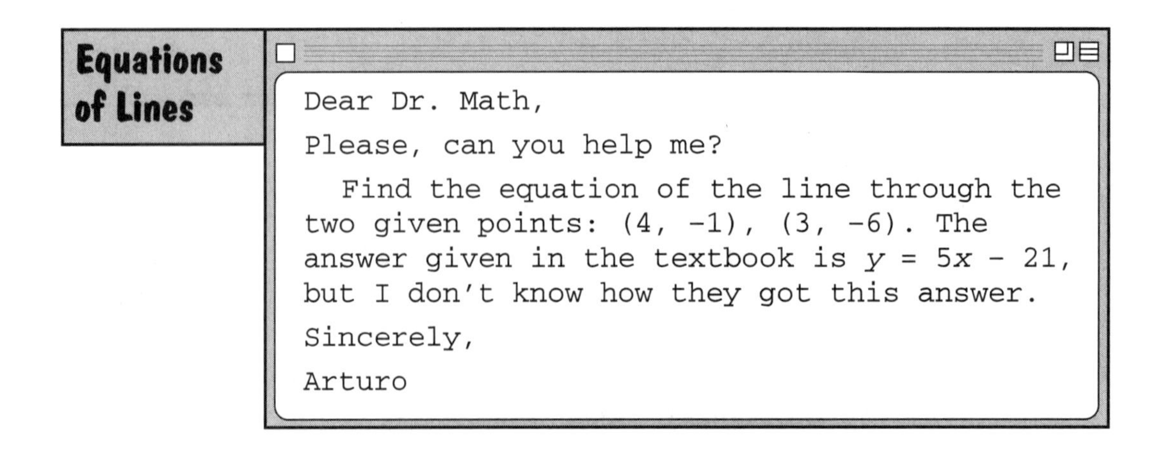

Equations of Lines

Dear Dr. Math,

Please, can you help me?

Find the equation of the line through the two given points: (4, −1), (3, −6). The answer given in the textbook is $y = 5x - 21$, but I don't know how they got this answer.

Sincerely,

Arturo

Dear Arturo,

You start by finding the slope of the line. This is a fraction where the numerator is the difference of the *y*-coordinates, and the denominator is the difference of the x-coordinates.

Make sure you take the differences in the same order, like this:

$$\frac{-1-(-6)}{4-3} = \frac{-1+6}{4-3} = \frac{5}{1} = 5$$

The slope is usually written as *m*, so here $m = 5$.

The form of the equation of a line you are asking about is called the slope-intercept form. This looks like $y = mx + b$ where *b* is the *y*-intercept, which is the *y*-value where the line intersects the *y*-axis. That is, $(0, b)$ is on the line.

What we know so far is that the equation looks like $y = 5x + b$. So, what is b? You can find that out by using either of the points that you know is on the line.

Let's try the first one, (4, –1). Since the combination of $x = 4$ and $y = -1$ must be on the line with equation $y = 5x + b$, it must be true that $-1 = 5(4) + b$. That is, –1 equals $20 + b$. The only b satisfying that is $b = -21$. So the equation is $y = 5x + (-21)$ or simply $y = 5x - 21$.

—Dr. Math, The Math Forum

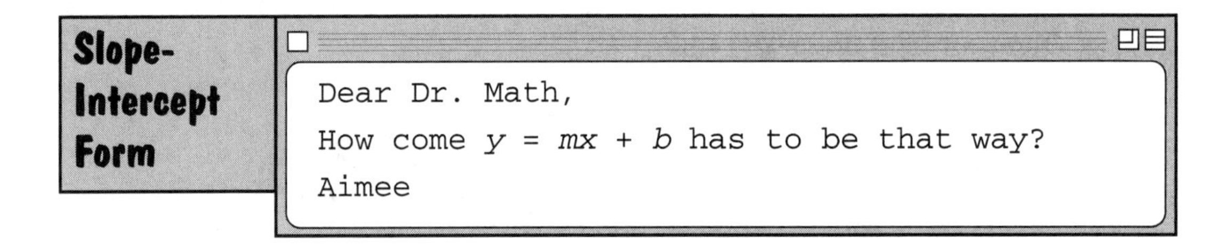

Slope-Intercept Form

Dear Dr. Math,

How come $y = mx + b$ has to be that way?

Aimee

Dear Aimee,

You can write any equation in a lot of different ways. What's nice about the form

$$y = mx + b$$

is that it allows you to sketch the graph of the corresponding line very quickly.

How do you do that? Well, you know that the line (if it's not vertical) has to cross the y-axis somewhere—in particular, where $x = 0$. Since you know $x = 0$,

$$\begin{aligned} y &= m(0) + b \\ &= b \end{aligned}$$

which tells you that $(0, b)$ is a point on the line. You also know that the line (if it's not horizontal) has to cross the x-axis—in particular, where $y = 0$. Since you know $y = 0$,

$$0 = mx + b$$

$$-b = mx$$

$$\frac{-b}{m} = x$$

which tells you that $(-\frac{b}{m}, 0)$ is a point on the line. So just by looking at the equation

$$y = mx + b$$

you can graph two points:

$$(0, b) \text{ and } (-\tfrac{b}{m}, 0)$$

And once you've graphed these two points, you can fill in the rest of the graph with a ruler.

Another nice thing about the form

$$y = mx + b$$

is that you can tell pretty quickly whether two lines are parallel or perpendicular. If they have the same slope and different intercepts, they are parallel. If the slopes can be multiplied to get –1, they are perpendicular.

For example, the following two lines are parallel:

$$y = 2x + 3$$

$$y = 2x + 14$$

And the following two lines are perpendicular:

$$y = 2x + 3$$

$$y = \frac{-1}{2}x - 6$$

—Dr. Math, The Math Forum

Writing in Slope-Intercept Form

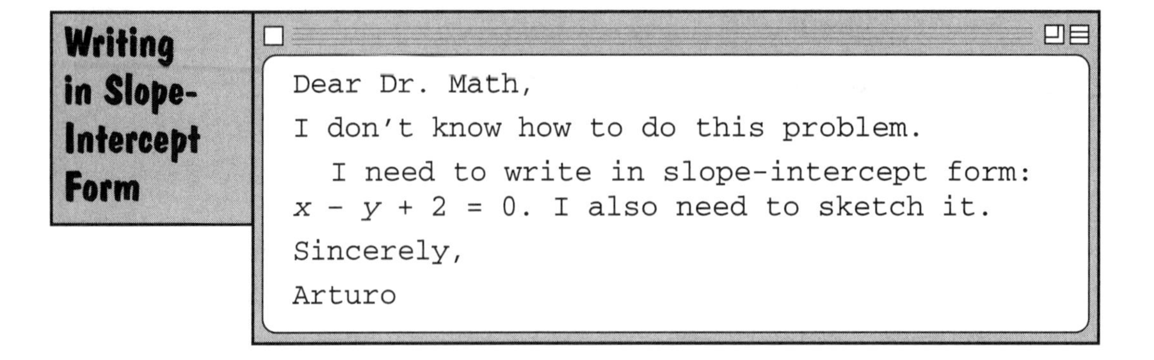

Dear Dr. Math,

I don't know how to do this problem.

I need to write in slope-intercept form: $x - y + 2 = 0$. I also need to sketch it.

Sincerely,

Arturo

Dear Arturo,

Slope-intercept form is great; it makes graphing much easier.

Let's take a line, say, $6x - 2y - 4 = 0$, and put it into slope-intercept form. Slope-intercept form is when a line is in the form

$$y = mx + b$$

where m is the slope and b is the y-intercept.

To get a line into that form, we just need to move terms around until it looks like that. So we start with the equation in the form given:

$$6x - 2y - 4 = 0$$

Now we move the x term over to the other side, so it looks like:

$$-2y - 4 = -6x$$

We move the 4 over to the other side too, because remember, we want the y term to be alone on one side. Now our equation looks like:

$$-2y = -6x + 4$$

We're almost there. The y term should have a **coefficient** of 1. (Remember a coefficient is a number that multiplies a variable—so we need to have something like $1y$ or y instead of $-2y$.) We need to divide both sides by -2 to get that, so the result looks like:

$$y = 3x + -2$$

The line is now in slope-intercept form. Now we need to graph it, and slope-intercept form makes it much easier. We know that if a line looks like $y = mx + b$, then m is the slope and b is the y-intercept. In the case of the line we're using, the slope is 3 and the y-intercept is -2.

The y-intercept is the place on the y-axis where the line intersects it. So, on your graph, plot the point $(0, -2)$, because we know that it lies on the line.

Slope can be defined in many ways, but one way of thinking about it is in terms of rise over run. Rise over run means that the slope is a ratio. In the case of this line, the slope is 3, and that can be expressed as $\frac{3}{1}$. That means for every 3 units you go up, you go 1 unit to the right.

Since you know one point already, you can find another point on the line by starting at that point and using the slope. Start at the point. Count up by the number from the top of the slope fraction, and over by the number from the bottom, and plot your next point. You could go over and then up too—the order doesn't matter as long as you keep the numbers straight! Here's what going over and then up with your slope of 3 from the point $(0, -2)$ looks like on a graph:

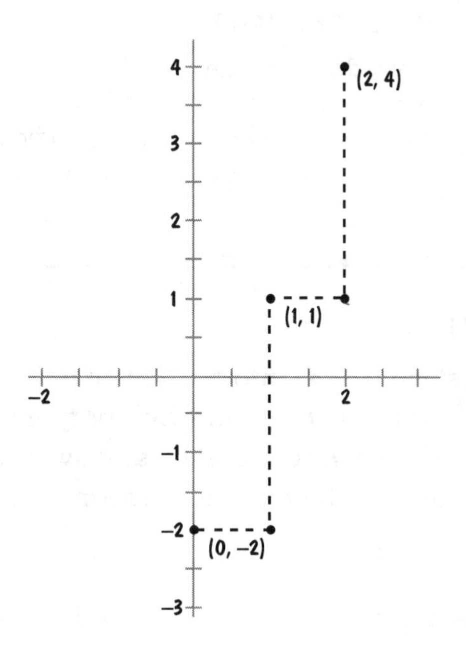

All of these points, (0, –2), (1, 1), (2, 4), and so on, are on the line $y = 3x - 2$. To draw the line, all you need to do is connect the points, and voilà! You're done.

You can use the same method I used to put $x - y + 2 = 0$ into slope-intercept form and then graph it. If you change your equation to slope-intercept form, you have:

$$y = x + 2$$

and so you know that the slope is 1 and the y-intercept is 2. If the y-intercept is 2, one point on your line is (0, 2). Here's the graph that shows that line:

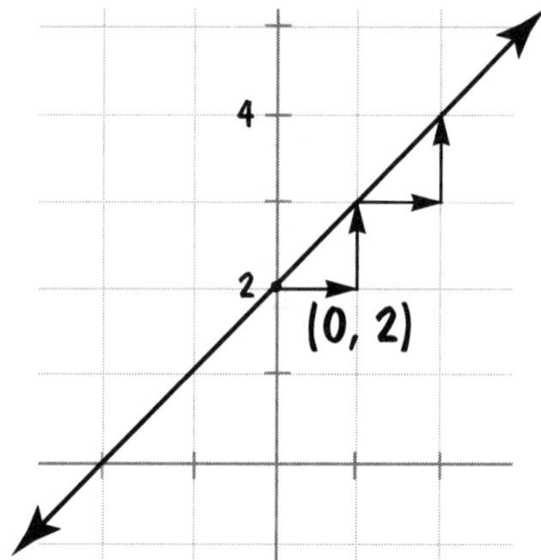

—***Dr. Math, The Math Forum***

Equations of Parallel and Perpendicular Lines

```
Dear Dr. Math,
How do you tell if the lines of these
equations are parallel, perpendicular, or
neither without graphing?
  4y - 5 = 3x + 1 and 12 = -6x + 8y - 3
Aimee
```

Dear Aimee,

The best way to get information on how two lines relate to each other (i.e., whether they are parallel, perpendicular, or neither) is to look at their slopes. And the easiest way to find the slopes is to get your equations into slope-intercept form, which means they look like this:

$$y = mx + b$$

The number represented by m in this equation is the slope.

I'll help you get one of your equations into the slope-intercept form:

You were given:	$4y - 5 = 3x + 1$
First add 5 to both sides:	$4y = 3x + 6$
Then divide both sides by 4:	$y = \frac{3}{4}x + \frac{6}{4}$

This means that the slope is $\frac{3}{4}$.

Try the other equation on your own and reduce the slope to the smallest possible fraction.

If the slopes of both equations are the same, it means the two lines are parallel. (This assumes, of course, that the y-intercepts are different. If they're the same *and* the slopes are the same, then the lines are the same!)

If one slope is the negative reciprocal of the other, then the two lines are perpendicular. To get the negative reciprocal of a number, put one over the number and then make it negative. So:

If the number is:	**Its negative reciprocal is:**
2	$-\frac{1}{2}$
$\frac{3}{10}$	$-\frac{10}{3}$
$-\frac{3}{2}$	$\frac{2}{3}$

So if your second equation had a slope of $-\frac{4}{3}$, then the lines of your equations would be perpendicular.

If the slopes are neither the same nor negative reciprocals, the lines are neither parallel nor perpendicular.

—*Dr. Math, The Math Forum*

Slope-Intercept and Point-Slope Forms

Dear Dr. Math,

My question isn't about how to use the point-slope form but more of what it is used for. Given a problem, I can accurately use: $y - y_1 = m(x - x_1)$ to get the answer, but I don't understand what its specific meaning is. Is it just another way of getting to a $y = mx + b$ style suitable for graphing, or does it serve some other purpose?

Sincerely,

Arturo

Dear Arturo,

Good question! Too often people learn forms but don't stop to ask what they are for.

There are several different ways to write the equation of a line, and each is designed to be used when you have certain pieces of information to start with. You could do everything with one form, such as slope-intercept, but often it's easier to use a form specially designed for one case. For example, there are:

Slope-intercept:	$y = mx + b$
Slope-x-intercept:	$y = m(x - a)$
Point-slope:	$y - y_1 = m(x - x_1)$
Two-point:	$\frac{y - y_1}{x - x_1} = \frac{y_2 - y_1}{x_2 - x_1}$
Two-intercept:	$\frac{x}{a} + \frac{y}{b} = 1$

These are all equivalent, and which one you use just depends on what you are given. But you don't have to memorize all of them. If you understand how graphs work, you can figure everything out when you need to. All these forms (except for the two-intercept form) are just ways of saying that the slope is constant. Either you are given the slope (m), or you figure out the slope $(y_2 - y_1)/(x_2 - x_1)$, and then you compare the general point (x, y) with either an intercept $(0, b)$ or $(a, 0)$, or another general point (x_1, y_1).

So the "meaning" of the point-slope form is simply that you can get the rise $(y - y_1)$ by multiplying the run $(x - x_1)$ by the given slope. You can graph it as it stands (by identifying the point and the slope), or you can change it into any other form you want. It's yours to use any way you like.

—Dr. Math, The Math Forum

Graphing Linear Equations

A graph of an equation is a picture of its solutions, all at once. In the case of linear equations, we can tell how big or small the slope of an equation is just by looking at the picture of the line it makes in the Cartesian plane.

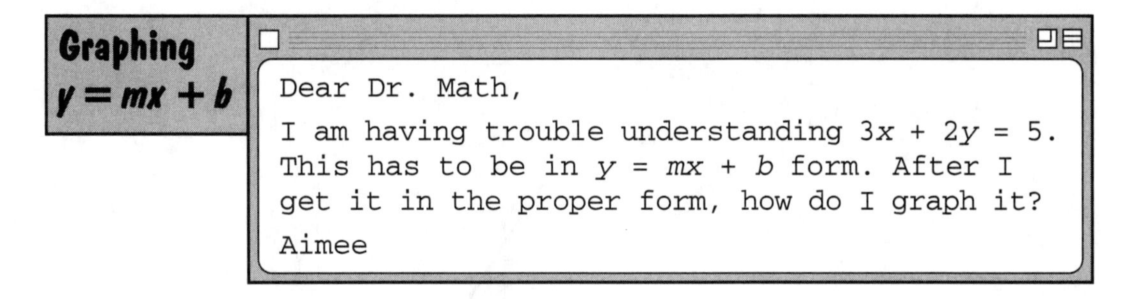

Graphing $y = mx + b$

Dear Dr. Math,

I am having trouble understanding $3x + 2y = 5$. This has to be in $y = mx + b$ form. After I get it in the proper form, how do I graph it?

Aimee

Dear Aimee,

Let's take a look at your question step-by-step. The first thing you told me is that you need to get the equation into $y = mx + b$ form. So, to begin, you will work on solving this equation for y. As you may recall, what that means is that you want to manipulate the equation so you get y by itself on one side of the equation. This is the process I would use:

$$3x + 2y = 5$$

$2y = -3x + 5$	Subtract 3x from each side to get 2y by itself.
$y = -\frac{3}{2}x + \frac{5}{2}$	Divide both sides by 2 to get y by itself.

Now that the equation is in the desired form, we're ready to graph it.

Remember that $y = mx + b$ is called "slope-intercept form." If you have an equation in this form, you will have the slope and the y-intercept for the graph. The slope is "m" (the coefficient of x) and is the rise over the run (or the change in y over the change in x). The y-intercept is "b" and is where the graph crosses the y-axis.

In this problem, $m = -\frac{3}{2}$ and $b = \frac{5}{2}$. I would first locate the y-intercept on the graph. Since $b = \frac{5}{2}$, I would find where $\frac{5}{2}$ (or $2\frac{1}{2}$) is on the y-axis and plot a point. This is where the graph crosses the y-axis.

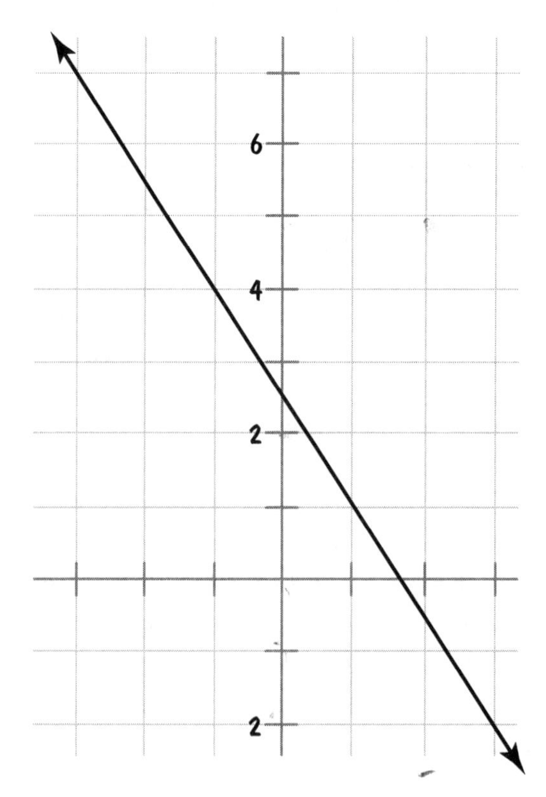

From there, I would use the slope to find other points on the graph. Slope is "rise over run." That means, in fraction form, the numerator indicates the distance you move in the vertical or *y* direction to find the next point, and the denominator indicates the distance you move in the horizontal or x direction. Keep in mind that a positive *y* direction is up, a negative *y* direction is down, a positive x direction is right, and a negative x direction is left. When you have a positive slope, you will use either a positive *y* and a positive x direction, or a negative *y* and a negative x direction. This is because a positive number divided by a positive number is positive, and a negative number divided by a negative number is also positive. Since we have a negative slope, we need one positive number and one negative number.

As I describe the graphing process below, I am assuming that you are using graph paper. This process will work with any scale on the graph (1 square could equal 2 units or 10 units, etc.), but I'm assuming 1 square equals 1 unit.

Our slope is $-\frac{3}{2}$. This means I can do one of two things: either go up 3 squares (positive direction) and left 2 squares (negative direction) or go down 3 squares (negative direction) and right 2 squares (positive direction). Be careful with this problem because you are beginning with a fraction (the $\frac{5}{2}$ from above)—you'll have to either estimate where halfway is or make each mark equal to $\frac{1}{2}$ (in other words, 2 marks on the graph = 1 unit).

Go to the $\frac{5}{2}$ you marked earlier. You can go up 3 squares from the $\frac{5}{2}$ and then move over left 2 squares and mark your next point. You can also go down 3 squares from the $\frac{5}{2}$ and then move over right 2 squares and mark your next point. If you do both of those operations, you will have three points and can connect the dots to make a line. (Although by definition, you need only two points in order to graph a line, adding a third point—or more—can be a good check that you're doing it correctly!) You can count out as many points as you need to make the line, but I suggest a minimum of three points for accuracy (although some teachers want you to use five points). The more points you have, the more accurate the line will appear on your graph.

—Dr. Math, The Math Forum

How to Draw Linear Equations

Dear Dr. Math,

I have a lot of trouble drawing linear equations. Under the $y = mx + b$ formula, say I had $y = (\frac{2}{3})x + 5$. How would I draw that? How would I go about finding other points?

Arturo

Dear Arturo,

Getting a mental picture of what an equation looks like is not always easy. But there are clues to help us.

The point-slope formula, $y = mx + b$, is used for straight lines, so anything that fits that equation is a straight line. Let's use what we know from the equation you were given to draw the line on a graph. The value of b in any point-slope equation is the y-intercept, and because $b = 5$ in your equation, that means the line crosses the y-axis at $y = 5$. So we will put a mark on our graph at 5 on the y-axis.

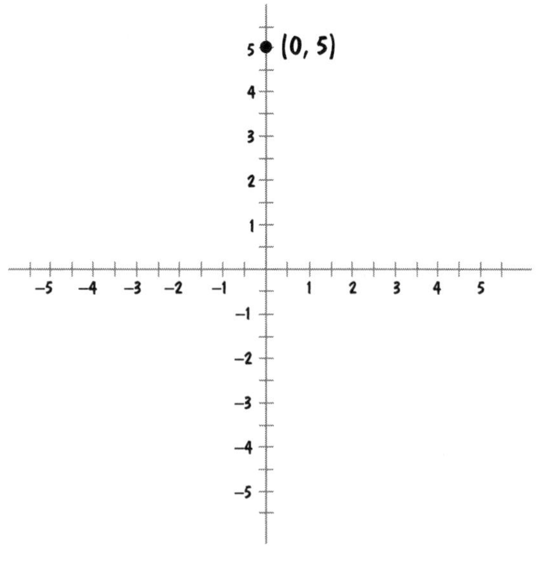

The value of m in any point-slope equation is the line's slope. In your equation, the slope is $\frac{2}{3}$. Slope is measured in rise over run. Since the slope is 2 over 3, then each step is a rise of 2 and a run of 3. So the second point will be 3 steps over to the right of and 2 steps above the first point.

Now that you have two points, you can take a ruler and draw a straight line that touches both points, put arrows on the ends, and the graph is done. Any other point that the line touches will also work in your equation. Try it out and see.

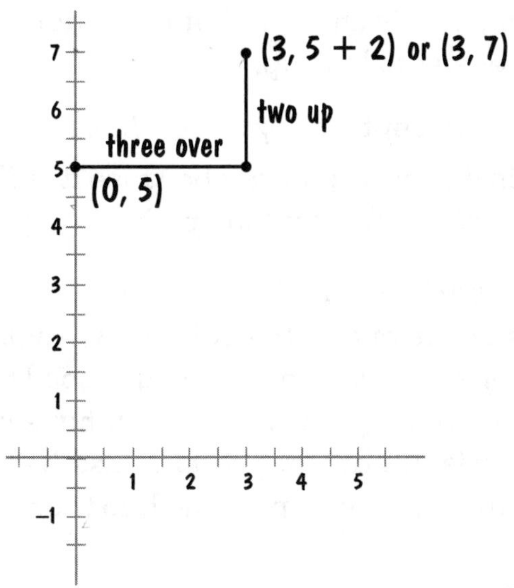

—*Dr. Math, The Math Forum*

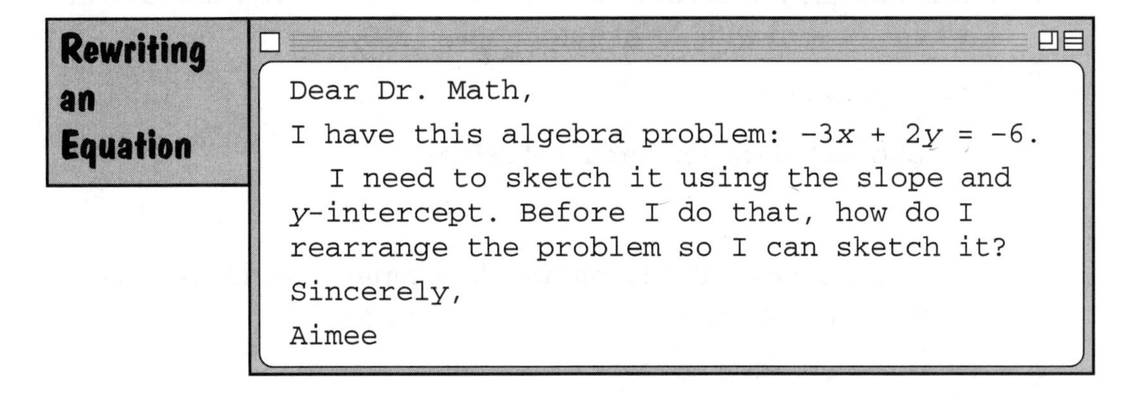

Dear Dr. Math,

I have this algebra problem: $-3x + 2y = -6$.

I need to sketch it using the slope and y-intercept. Before I do that, how do I rearrange the problem so I can sketch it?

Sincerely,

Aimee

Dear Aimee,

Linear equations with two variables, such as the one you've asked about, can be written in a number of ways. The best way to write a linear equation depends on the information you want to get out of it. Here are three types:

1. General form: $ax + by = c$.

   This is the form of the equation you've asked about. Writing equations in this form allows them to be ordered with other more complicated equations—in much the same way that putting words in alphabetical order in a dictionary makes them easier to look up.

2. Slope-intercept form: $y = mx + b$.

   This is the form you want because it will let you easily see the slope, m, and the y-intercept, b.

3. Point-slope form: $y - y_1 = m(x - x_1)$.

   The slope is m and the point is $(x_1 - y_1)$. This is a variation of the slope-intercept form, which is good to use if instead of having the intercept point $(0, b)$, you have a specific point not on the y-axis. It's an easy form to use because you can plot the point and use the slope of the line to draw the line through that point.

To get from the general form to the slope-intercept form, you need to isolate the y term on one side of the equation. Here's how to do this when you start with the general form:

$$ax + by = c$$

Subtract the ax term from both sides:

$$by = c - ax$$

Rewrite the right side, putting the x terms first (this isn't necessary, but it looks better):

$$by = -ax + c$$

Divide both sides by b to get the y alone on one side:

$$\frac{by}{b} = \frac{-ax}{b} + \frac{c}{b}$$

$$y = \frac{-ax}{b} + \frac{c}{b}$$

Now you have an equation in the slope-intercept form, so you can see that the slope is $-\frac{a}{b}$ and the y-intercept is $\frac{c}{b}$.

Use this procedure to rewrite your equation, and you should be all set.

—Dr. Math, The Math Forum

Resources on the Web

Learn more about linear equations at these sites:

Math Forum: Chameleon Graphing: Lines and Slope

mathforum.org/cgraph/cslope/

A Web unit for middle school and early high school students, in which Joan the Chameleon introduces and explores lines and slope.

Math Forum: Graphing Linear Functions

mathforum.org/alejandre/linear.graph.html

Step-by-step directions to create graphs of linear functions using a ClarisWorks spreadsheet file.

Math Forum: Middle School Algebra Links

mathforum.org/alejandre/frisbie/math/algebra.html

A variety of resources on the Web emphasizing algebraic thought and addressing each of the particular NCTM standards for grades 6, 7, and 8.

Shodor Organization: Project Interactivate: Slope Slider

shodor.org/interactivate/activities/slopeslider/

This activity allows the manipulation of a linear function of the form $f(x) = mx + b$ and encourages the user to explore the relationship between slope and intercept in the Cartesian coordinate system.

Shodor Organization: Project Interactivate: Linear Function Machine

shodor.org/interactivate/activities/lfm/

Students investigate linear functions by trying to guess the slope and intercept from inputs and outputs.

Shodor Organization: Project Interactivate: Positive Linear Function Machine

shodor.org/interactivate/activities/plfm/

Students investigate linear functions with positive slopes by trying to guess the slope and intercept from inputs and outputs.

Systems of Equations

A system of equations is a number of equations that are all true together. (The terms "system of equations" and **"simultaneous equations"** are used interchangeably.) This means that you don't just find the answer to the first equation and then the answers to the others. Instead you find an answer or answers that work for all of the equations at the same time. The case of two linear equations with two **unknowns** is the easiest case to learn about first. When we find the

values of the variables, say x and y, that make both equations true at the same time, then we've found the solution of this system of equations. There may be many pairs of x and y that make the first equation true, and many pairs of x and y that make the second equation true, but we are looking for an x and a y that will work in both equations at the same time.

In this part, Dr. Math explains:

- simultaneous equations
- graphing simultaneous equations
- inconsistent and dependent systems

1 Simultaneous Equations

When two or more conditions have to share a solution, you have to solve equations *simultaneously,* or as a *system,* instead of individually. This can be true of many kinds of equations, but linear equations are the easiest to deal with, so we'll talk about those in this part. But the principles remain the same for all kinds of equations.

There are several ways to go about solving a pair of linear equations. The most popular ones are: inspection, **elimination** (using addition or subtraction of equations to eliminate one variable), **substitution** (replacing one variable with its value in terms of the other), **intersection** (solving both equations for one value and setting the resulting expression equal to each other), and graphing.

This section will give you an overview of simultaneous equations and introduce you to some typical techniques for solving them.

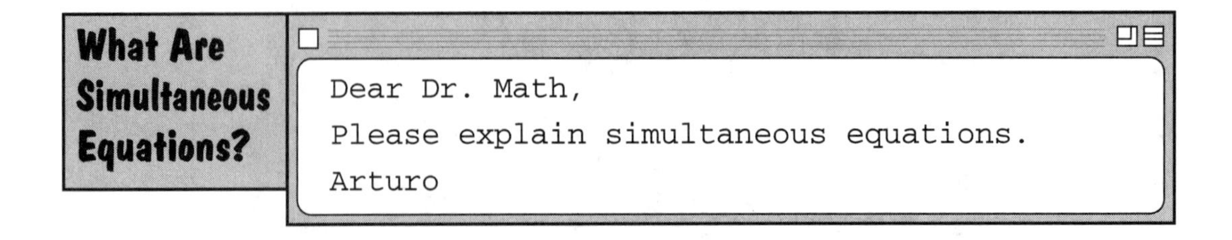

Dear Arturo,

When you have an equation like:

$$3x + 2y = 5$$

there are infinitely many pairs of values for x and y that can make the equation true. For example,

$$3(1) + 2(1) = 5$$

so ($x = 1, y = 1$) is a solution to the equation. Also,

$$3\left(\frac{1}{3}\right) + 2(2) = 5$$

so ($x = \frac{1}{3}, y = 2$) is also a solution.

An equation like this is a little like the volume control on a stereo. As you move the control to different positions (change the value of x), you increase or decrease the volume of the music (increase or decrease the value of y).

But what if you have more than one equation, and (both or) all of them have to be true at the same time? For example, let's say you have the equations:

$$3x + 2y = 5$$

and

$$5x + y = 13$$

If we consider either one of these equations by itself, we can see that it has an infinite number of possible solutions. But what if we consider them together? Then it turns out that there is only one way to make *both* equations true with a single pair of values for x and y.

This can seem very mysterious, until you think about what's really happening here. Each of these equations describes a line. Each point on a line corresponds to a solution of the equation of the line. And unless two lines are parallel, they intersect at exactly one point. This point is the solution of the simultaneous equations.

—***Dr. Math, The Math Forum***

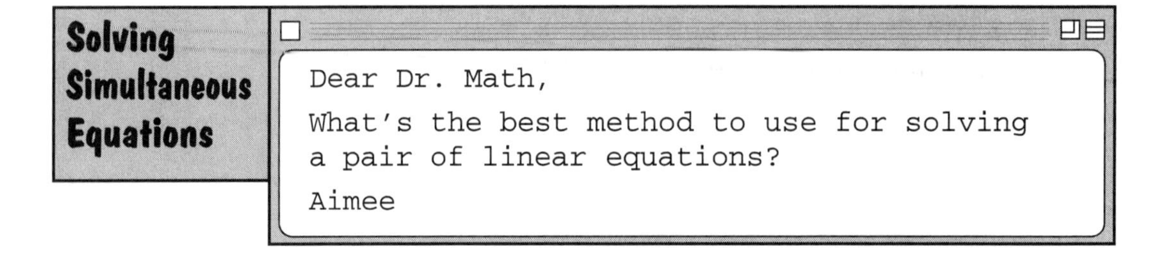

Solving Simultaneous Equations

Dear Dr. Math,

What's the best method to use for solving a pair of linear equations?

Aimee

Hi, Aimee,

The best method is the one that requires the least work! However, that will depend on the particular equations that you're trying to solve. The most popular methods are: inspection (just looking at them!), elimination, substitution, intersection (solving for one variable and setting the resulting expressions equal to each other), and graphing.

Graphing is usually the hardest, unless you have a graphing calculator or program handy. Then it can be the easiest—but there is always room for error in trying to read a point of intersection from a graph. So graphing is good when you just need an approximate answer. If you need a lot of precision, you should probably stick with one of the other methods.

Inspection is the easiest method, and it's always a good idea to check to see if you can use it before jumping in with another method. When can you use inspection? It's easiest when one of the variables has the same coefficient in both equations, for example:

$$3x + 2y = 12$$
$$3x + 5y = 18$$

To use inspection, you reason this way: "In moving from the first equation to the second, all that changes is that we're adding $3y$ on the left side. On the right side, we add 6. So it must be the case that $3y$ is the same as 6, which means that y must equal 2."

In fact, all you're really doing here is using elimination, but you're not bothering to do the formal addition or subtraction of equations. If we write the equations in the opposite order,

$$3x + 5y = 18$$
$$3x + 2y = 12$$

we can subtract the left sides and the right sides to get:

$$(3x - 3x) + (5y - 2y) = (18 - 12)$$
$$3y = 6$$

which is what we did the last time but using insight instead of equations. The nice thing about elimination is that it continues to work even when inspection fails, which is to say, when all the coefficients are different:

$$3x + 2y = 12$$
$$6x + 5y = 26$$

To make elimination work in a situation like this, you need to multiply one of the equations by a constant factor so that you end up with matching coefficients. In this case, we can multiply the first equation by 2 to get:

$$6x + 4y = 24$$
$$6x + 5y = 26$$

Now we can use inspection or elimination, depending on how much we trust ourselves to do the work in our heads instead of on paper!

That case was sort of like adding $\frac{1}{2}$ and $\frac{1}{4}$, where one of the denominators is already the one you're going to use as a common denominator. But sometimes you get equations like:

$$3x + 2y = 12$$
$$7x + 5y = 26$$

And this is where elimination starts to seem like more trouble than it's worth, because you have to multiply *both* equations by different factors to get matching coefficients:

$$7(3x + 2y) = 7(12)$$
$$5(7x + 5y) = 5(26)$$

At this point, other methods start to look pretty good! The remaining methods—substitution and intersection—both use the same first step, which is to select one of the equations and solve it for one of the variables in terms of the other. In the example above, we might choose the first equation and do this:

$$3x + 2y = 12$$

$$2y = 12 - 3x$$

$$y = 6 - \frac{3}{2}x$$

At this point, if we want to use substitution, we go back to the other equation, locate every occurrence of y, and replace it with the expression $(6 - \frac{3}{2}x)$:

$$26 = 7x + 5y$$

$$= 7x + 5\left(6 - \frac{3}{2}x\right)$$

Now we're in familiar territory again: one equation, with one variable. We can solve this to find the value of x, and plug that into either equation to find the value of y.

In intersection, we reason a little differently. Remembering that we're dealing with lines, we think: "Assuming the lines aren't parallel, they have to intersect *somewhere,* and wherever that happens, they have to have the same value of y (or of x)." So we solve *both* equations for the same variable, and set the expressions equal to each other:

$3x + 2y = 12$ becomes $y = \frac{12 - 3x}{2}$

$7x + 5y = 26$ becomes $y = \frac{26 - 7x}{5}$

and since it must be true that $y = y$, it's true that

$$\frac{12 - 3x}{2} = \frac{26 - 7x}{5}$$

And again, we're in familiar territory.

Personally, I prefer elimination to substitution and intersection, because I like to work with integers whenever possible. Of course, if I'm starting out with coefficients that aren't integers, then that excuse goes out the window, and I'll usually reach for intersection. (I prefer intersection to substitution, because I think the equations tend to look a little nicer.)

Of course, everyone is different, which is why we have all these methods in the first place! At the beginning of this message, I said that the best method is the one that requires the least work. I guess I'd say that the amount of work is only the second most important consideration. The *best* method is the one that you have the most confidence in. (As they say, you can't make mistakes fast enough to get a correct answer!) And that's something that you can decide only after you've had some practice with all the different methods.

—*Dr. Math, The Math Forum*

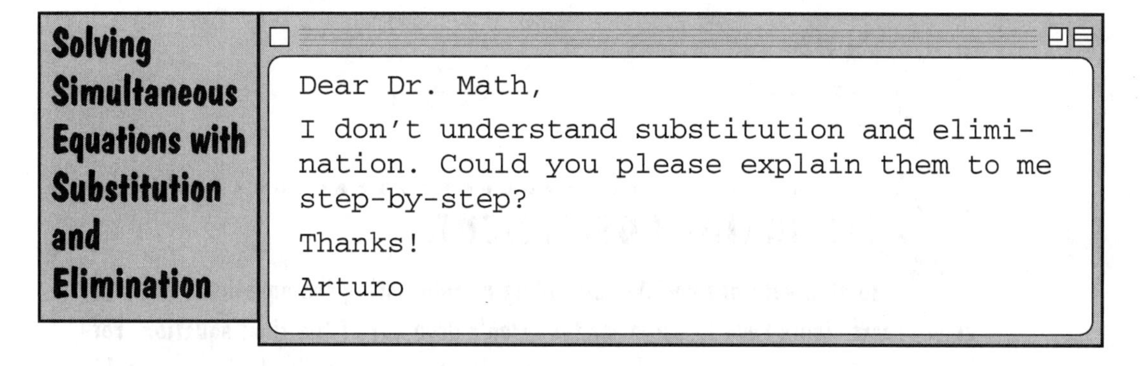

Solving Simultaneous Equations with Substitution and Elimination

Dear Dr. Math,

I don't understand substitution and elimination. Could you please explain them to me step-by-step?

Thanks!

Arturo

Hi, Arturo,

Suppose I have these two equations:

$$3x + y = 9$$
$$4x - y = 5$$

Elimination works by adding or subtracting equations to get rid of (or eliminate) one of the variables. In this case, I can add the two equations:

$$(3x + y) + (4x - y) = 9 + 5$$

Do you see why this works? This is really just an application of the basic rule in algebra, which is that I can add the same thing to both sides of any equation. That is,

$$(3x + y) + [\text{something}] = 9 + [\text{the same thing}]$$

After adding the equations, the y conveniently disappears, and I can solve for x.

$$\begin{aligned} 3x + y + 4x - y &= 9 + 5 \\ 3x + 4x + y - y &= 14 \\ 7x &= 14 \\ x &= 2 \end{aligned}$$

Once I know the value of x, I can plug it back into either equation to get the corresponding value of y.

Now, what if things didn't work out so conveniently? What if I'd had these equations instead?

$$\begin{aligned} 3x + 2y &= 12 \\ 4x - y &= 5 \end{aligned}$$

Rx MATCHING COEFFICIENTS

To eliminate one variable by adding or subtracting the equations, the variables' coefficients have to agree, or they won't drop out of the final equation. For instance, if I added the original second equation to the first, "as is," there would still be one y in the final equation:

$$\begin{aligned} 3x + 2y + (4x - y) &= 12 + 5 \\ 3x + 4x + 2y - y &= 17 \\ 7x + y &= 17 \end{aligned}$$

That doesn't help one bit. But what if we added the second equation twice? Then another y would get subtracted from the total, and we'd have none left, which is what we want. Adding the whole equation twice is the same as multiplying it by 2 and then adding. That's because the coefficients of the y variables agree.

Well, in this case, I can still use elimination, but I'd have to scale one of the equations to get back to a situation where addition or subtraction would work. **Scaling** means making the components of an equation bigger or smaller but keeping the same relationships between the variables, like making a *scale* model. In this case, I can multiply everything in the second equation by 2 (do you see why this works?) to get:

$$3x + 2y = 12$$
$$8x - 2y = 10$$

and now I'm more or less back where I was before. When I add the equations, the y disappears so I can solve for *x:*

$$3x + 2y + 8x - 2y = 12 + 10$$
$$3x + 8x + 2y - 2y = 22$$
$$11x = 22$$
$$x = 2$$

Sometimes it may be necessary to scale both equations, for example,

$$3x + 2y = ?$$
$$5x + 3y = ?$$

Here, you'd want to multiply the first equation by 3, and the second by 2, to get rid of the *y* terms; or you could multiply the first equation by 5, and the second by 3, to get rid of the x terms. It's very similar to the process of finding a common denominator in order to add fractions. The main thing to remember is that you need to have an identical term (for example, $2y$ in your original equations) in both equations, so that you can get the variable in that term to disappear.

Sometimes you may have to subtract one equation from the other, rather than adding them. For example, if you have these two equations:

$$3x - 2y = 12$$
$$8x - 2y = 10$$

and you subtract them, the *y* disappears so you can solve for *x:*

$$(3x - 2y) - (8x - 2y) = 12 - 10$$
$$3x - 2y - 8x + 2y = 2$$
$$3x - 8x - 2y + 2y = 2$$
$$-5x = 2$$
$$x = -\frac{2}{5}$$

What about substitution? For that method, you solve one of the equations for one variable in terms of the other and then substitute that into the other equation. Let's look at these equations again:

$$3x + 2y = 12$$
$$4x - y = 5$$

I can rewrite either equation so that the y is on one side:

$$3x + 2y = 12 \qquad\qquad 4x - y = 5$$
$$2y = 12 - 3x \qquad\qquad 4x = 5 + y$$
$$y = 6 - \frac{3}{2}x \qquad\qquad 4x - 5 = y$$

Or I could rewrite them so that the x is alone on one side. It doesn't really matter. The important thing is that I now have an expression for y in terms of x (or an expression for x in terms of y). So wherever I see the variable, I can substitute the expression—for example, in the other equation.

$$3x + 2y = 12$$
$$3x + 2(4x - 5) = 12$$
$$3x + 8x - 10 = 12$$
$$11x = 22$$
$$x = 2$$

Since I know that $y = 4x - 5$, I can substitute for y and solve for x

As before, I can now use the value of x to get the value of y by plugging it into either equation.

—Dr. Math, The Math Forum

Solving Equations in Two Variables

Dear Dr. Math,

The problem is

$$2x + 3y = 7$$
$$4x - 5y = 25$$

I have no idea how to solve this problem at all. Please help me; I'm really stuck.

Aimee

Dear Aimee,

Suppose for a moment that we are trying to solve a slightly different problem:

$$2x + 3y = 7$$
$$2x + 4y = 10$$

Just looking at the two equations, we can see that an extra y on the left side corresponds to an extra 3 on the right side. So we know that $y = 3$ without having to do anything.

If instead we have:

$$2x + 3y = 7$$
$$2x - 4y = 21$$

We see that a loss of $7y$ on the left (from $+3y$ to $-4y$) corresponds to a gain of 14 on the right. So we know that $-7y = 14$, or $y = -2$.

By now, you might be wishing that you had to solve a problem like this instead of the one that you were given. But wait! Is there a way that you can turn your problem into a problem like this?

What if you multiply everything in the first equation by 2? Then you would have:

$$4x + 6y = 14$$
$$4x - 5y = 25$$

Can you take it from here?

—Dr. Math, The Math Forum

Find the Solution Set

Dear Dr. Math,

I have a question. Solve the system:

$$3x - 6y = 1$$
$$x = 2y + 3$$

Thanks for all your help. I couldn't do this without all of you.

Sincerely,

Arturo

Dear Arturo,

I'll try doing a problem similar to yours so you'll have a chance to practice on your own problem.

Your problem is a pair of simultaneous equations with a little twist. I'll show you how to solve another pair of equations, which has the same twist:

$$3x - 9y = 3$$
$$x = 3y + 4$$

First you have to put them in the same form, by moving the $3y$ to the left:

$$3x - 9y = 3$$
$$x - 3y = 4$$

There are several ways to solve these types of problems. I'll use the elimination method by making the coefficients of x the same and subtracting the resulting equations. To do this, multiply the second equation by 3:

$$3x - 9y = 3$$
$$3x - 9y = 12$$

You would normally subtract these, but if you did, you would get $0x + 0y = -9$, which can never be true. How can that be? It means that these two equations represent lines that never intersect, because they are parallel. Whenever you end up with an impossible equation, it means there are no solutions.

Now we've shown that there are no solutions using the *elimination* method. We can show the same thing by using the *substitution* method. To do that, we substitute the expression $(3y + 4)$ into the first equation wherever we see an x:

$$3x - 9y = 3$$
$$3(3y + 4) - 9y = 3$$
$$9y + 12 - 9y = 3$$
$$12 = 3$$

Again, this is an impossible equation, which means that there are no solutions. Finally, note that we could have arrived at the same conclusion by putting the first equation in the same form as the second:

$$\begin{aligned} 3x - 9y &= 3 \\ 3x &= 9y + 3 \\ x &= 3y + 1 \end{aligned}$$

By thinking about it a little, we can see that there is no value that you can choose for y that gives you the same value of x in both equations! But we can show that directly by setting our two expressions to be equal to each other:

$$\begin{aligned} x &= x \\ 3y + 4 &= 3y + 1 \\ 4 &= 1 \end{aligned}$$

As I said earlier, if you end up with an impossible equation, it means there are no solutions. All of the methods that you would normally use to solve a system of equations should bring you to the same conclusion, albeit in slightly different forms.

—Dr. Math, The Math Forum

2 Graphing Simultaneous Equations

Another way to find the solution of two simultaneous equations is to graph two equations on the same graph and see where the lines intersect. The point of intersection will give you the x and y values that satisfy both equations. If the two lines on the graph are parallel, then there is no solution to your simultaneous equations.

How to Graph Systems of Equations

Dear Dr. Math,

Solve the following system of equations by graphing

$$3x + 2y = 5$$
$$-3x - 2y = 10$$

Sincerely,
Aimee

Dear Aimee,

Do you know how to graph a linear relation between x and y? You can always start with a value for x and then determine the value of y required to satisfy the equation.

For example, use the equation:

$$3x + 2y = 5$$

If x is 0, then y must be $\frac{5}{2}$. If x is 1, then y must be 1. Since two points define a line, all you have to do is draw the line that connects $(0, \frac{5}{2})$ to (1, 1).

The solution to a system of linear equations is the values of x and y at the point at which the lines intersect.

Don't be upset that you can't solve this system you were given, because there is no solution. If you multiply both sides of the second equation by –1, you'll see $3x + 2y$ can't be 5 and –10 at the same time.

If you graph these two lines, you'll see that they are parallel.

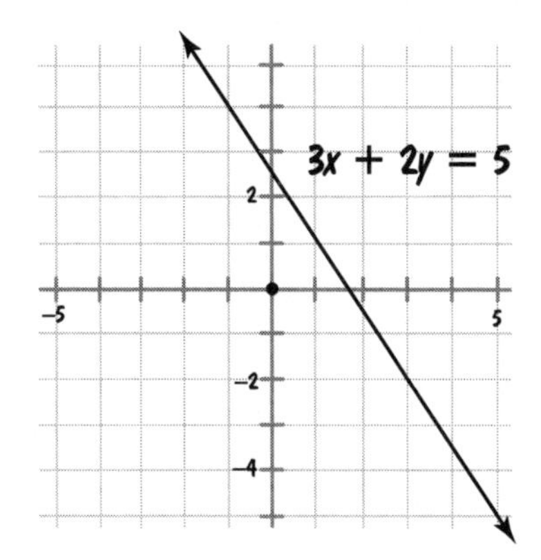

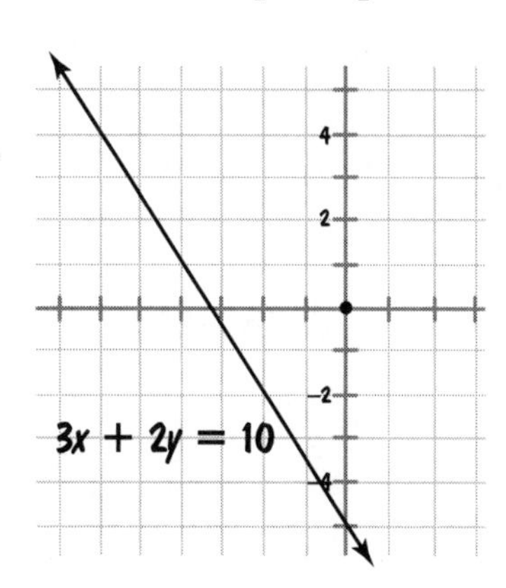

You can also see this without graphing, by re-expressing each equation in the slope-intercept form of a linear function in which $y = mx + b$ where m is the slope of the line and b is the y-intercept. Your two equations are:

$$3x + 2y = 5$$
$$2y = -3x + 5$$
$$y = -\frac{3}{2}x + \frac{5}{2}$$

and

$$-3x - 2y = 10$$
$$-2y = 3x + 10$$
$$y = -\frac{3}{2}x - 5$$

With your equations in this form, you can see that the two lines have the same slope, but they have different y-intercepts; therefore they must be parallel.

—Dr. Math, The Math Forum

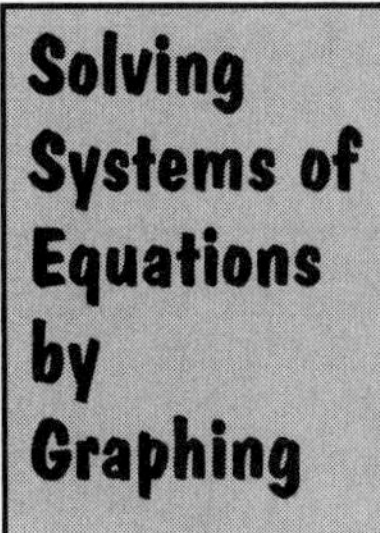

Solving Systems of Equations by Graphing

Dear Dr. Math,

I need to learn how to identify solutions and how to find solutions by graphing. For example, I need to solve by graphing

$$x + 2y = 7$$
$$x = y + 4$$

Arturo

Dear Arturo,

For the first equation, pick a few values of y and compute the value of x that corresponds:

$y = 0, x = 7$

$y = 1, x = 5$

$y = 3, x = 1$

$y = 4, x = -1$

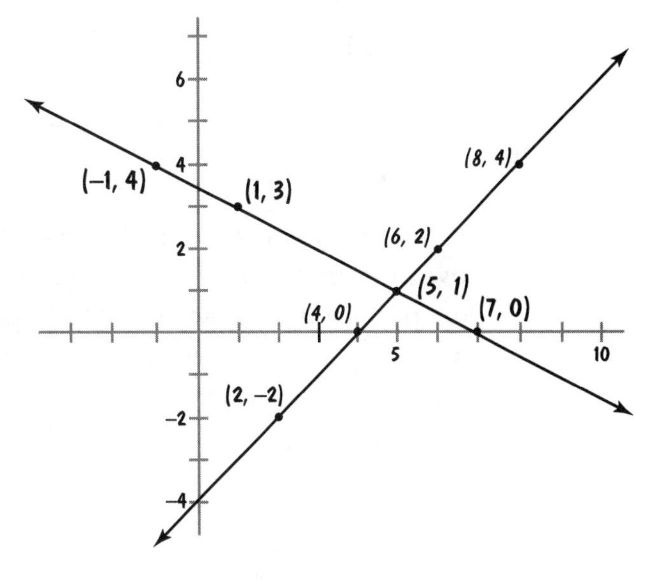

Now plot the points (7, 0), (5, 1), (1, 3), and (–1, 4), and draw a line through them.

Do the same for the second equation: (2, –2), (4, 0), (6, 2), and (8, 4).

Now see if you can figure out the coordinates of the point where the two lines intersect.

—Dr. Math, The Math Forum

Inconsistent and Dependent Systems

Systems of equations don't always work out perfectly. When we can't find an answer, it's possible that we have a system that has no solution. We call these cases **inconsistent** or **overdetermined** systems. Another possibility is when there is more than one solution. In that case the system is called **dependent** or **underdetermined.**

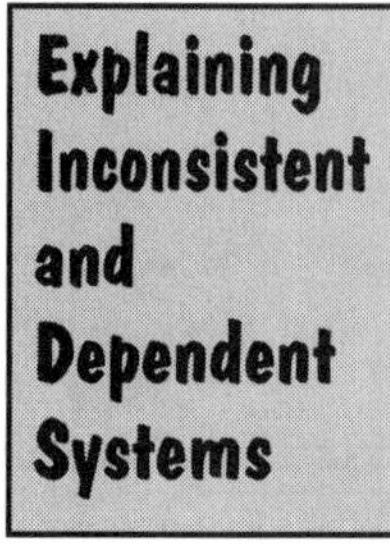

Dear Dr. Math,

Write a second linear equation for $y = 2x + 4$ that would create an inconsistent system. Also write a second linear equation that would create a dependent system.

I have definitions of both, but I don't know how to do the problem.

Sincerely,

Aimee

Dear Aimee,

If you've got the definitions, you know that an inconsistent (or overdetermined) system is one without a solution, and a dependent (or underdetermined) system is one where there is more than one solution. Here are some simple examples of inconsistent or overdetermined systems:

$0 = 1$

$y = 1$ and $y = 2$

$x = y$ and $x = y + 1$

Now, some comments on why each of these is inconsistent. The equation $0 = 1$ is a false statement, and even though it is not strictly a system, or even a statement about unknown quantities (like x or y), it is inconsistent because it is false! The equations $y = 1$ and $y = 2$

are a system, but it cannot be true, because it implies that $1 = 2$; $x = y$ and $x = y + 1$ are a similar situation because this system implies that $0 = 1$.

Look at these examples of dependent or underdetermined systems:

$$x = y$$

$$y = 3x \text{ and } 2y = 6x$$

$$x + 2y = 0$$

Here's why each of these systems is dependent. In $x = y$, there are infinitely many solutions, like $x = y = 5$ or $x = y = -1$. In $y = 3x$ and $2y = 6x$, the second equation is just the same as the first when divided by 2, so it provides no additional information about x or y. Finally, $x + 2y = 0$ is the same as saying $x = -2y$, which has infinitely many solutions, like $x = -2$ and $y = 1$ or $x = 4$ and $y = -2$.

Now say you are given $y = 3x + 5$. You want to write another equation that, with the given equation, makes an inconsistent system. There are lots of ways to do this; one way is to say

$$y = 3x + 6$$

Which would mean that $3x + 5 = 3x + 6$, or $5 = 6$. Another option is to just use

$$y = 3x$$

Now suppose you want to make a dependent system out of the same original equation. One way would be to multiply both sides by some number.

$$2y = 6x + 10$$

You could also do it another way, by adding or subtracting some number from both sides. Let's subtract 4 from both sides:

$$2y - 4 = 6x + 6$$

Now you can see some ways to turn your equation into independent and dependent systems.

—Dr. Math, The Math Forum

Finding Types of Linear Systems by Graphing

Dear Dr. Math,

In my math class, I was told to find out what is meant when a type of linear system is said to be a consistent or an inconsistent system.

My friend and I have looked up this question in a few math books and have asked three or four teachers without any success.

Can you help?

Arturo

Dear Arturo,

When you are talking about linear systems, it is helpful to graph the equations that make up the system and see how the resulting lines relate to one another.

If the system consists of two equations, there are three possibilities when you graph the equations:

1. The lines may cross at *one* point. This point is usually called the solution of the system. A system with one solution is called **consistent**.
2. The lines may be **coincident** (lie on each other). There are infinitely many solutions to the system. In fact, any point that lies on the line is a solution. This type of system is called dependent.
3. The lines may be parallel. There are *no* solutions to the system. This type of system is called **inconsistent.** Note that we can tell if a system is inconsistent without actually graphing it by looking at the slopes of the lines and their y-intercepts. If the slopes are the same but the y-intercepts are different, the lines are parallel and not the same line.

– Dr. Math, The Math Forum

Resources on the Web

Learn more about systems of equations at these Math Forum sites:

Algebraic Problem-Solving Using Spreadsheets

mathforum.org/workshops/sum98/participants/sinclair/problem/intro.html

A unit that emphasizes moving from numerical to symbolic representation.

Investigating Functions with Excel

mathforum.org/workshops/sum98/participants/sinclair/function.html

An investigation of functions using spreadsheets.

Middle School Problem of the Week: Mirror Musing

mathforum.org/midpow/solutions/solution.ehtml?puzzle=147

Using the information given in a picture of Hagrid and Baby Norbert, find out how tall each of them is.

Problems Library: Algebra: Functions and/or Graphing

mathforum.org/library/problems/sets/alg_functions.html

A variety of Algebra Problems of the Week that involve functions and/or graphing.

Problems Library: Algebra: Linear Equations

mathforum.org/library/problems/sets/alg_linear.html

A variety of Algebra Problems of the Week that involve linear equations.

Problems Library: Algebra: Systems

mathforum.org/library/problems/sets/alg_systems.html

A variety of Algebra Problems of the Week that involve systems of equations.

Polynomials

The word **polynomial** comes from the Greek *poly-*, meaning "many," and Latin *nomen,* meaning "name" or "named thing"—a polynomial, then, is a bunch of things, or mathematically, two or more terms, strung together.

Each term includes a variable, raised to some nonnegative integer power, sometimes with a coefficient in front of it. Sometimes there's a term with no visible variable, called a **constant,** but really

it does have a variable, raised to the 0th power! In the example below, 11 is the constant, and there are five terms (including the constant).

$$7x^5 + x^4 - 4x^3 + 15x^2 + 11$$

The term raised to the highest power will indicate the **degree** of the polynomial. The above polynomial has a degree of 5.

So, what's so special about polynomials? Mostly, they're easy to work with. They have a nice, compact notation, with no special cases. They're well-behaved. If you add enough terms, you can use them to approximate other functions that aren't so easy to work with—for example, the path of a spacecraft through the solar system, or the shape of the sound wave that results when you pluck a guitar string.

In this part, Dr. Math explains:

- monomials and polynomials
- simplifying polynomials
- adding and subtracting polynomials
- multiplying polynomials
- patterns in polynomials
- dividing polynomials

1 Monomials and Polynomials

A **monomial** is a polynomial with only one term. Formally, a monomial is an expression of the form ax^n where a is any number and n is a positive integer. The number a (which could be 0) is the coefficient of the monomial, and the integer n is the degree of the monomial. A polynomial is at least two monomials strung together by addition or subtraction. This can be formally represented as $ax^0 + ax^1 + ax^2 + \ldots + ax^n$.

Besides thinking of polynomials just in terms of algebraic symbols, take a look at this geometric representation of the polynomial $x^2 + 4x + 3$:

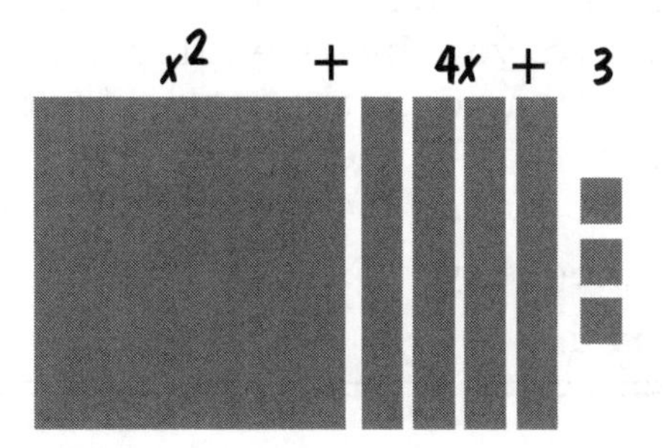

Of course, you can't do this with all polynomials, but when you are working with polynomials of degree 2 (where the highest power of any term is 2), geometric pictures can help you remember what the algebraic notation represents.

There are three parts to this picture. Each part corresponds to one of the terms of the polynomial. First there is one large square with a side length of x. We write that as x^2. Next there are four rectangles each measuring 1 unit by x units. So, we have 4x. The three small squares have a side length of 1 unit, and so we have 3. When we put that all together, we have the polynomial $x^2 + 4x + 3$.

GEOMETRIC REPRESENTATION

If we think of the pictured example given above, we can see how the large square, four rectangles, and three small squares can be rearranged to make a large rectangle that has one side of length $(x + 3)$ and another side of length $(x + 1)$.

$$x^2 + 4x + 3 = (x + 3)(x + 1)$$

Monomials

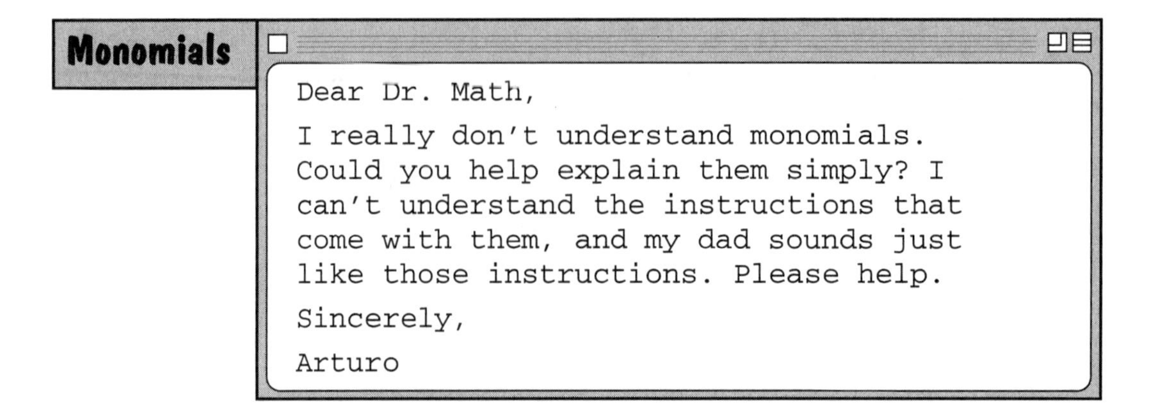

Dear Dr. Math,

I really don't understand monomials. Could you help explain them simply? I can't understand the instructions that come with them, and my dad sounds just like those instructions. Please help.

Sincerely,

Arturo

Dear Arturo,

There's actually not all that much to know about monomials.

The way you make a monomial is to take one or more variables (like x or g or r) and/or constants (like 0 or 1 or 5 or 247) and multiply them together. These are also called "terms." Here are some examples of monomials:

x	$5b^2$
$3y$	$3xy$
8	

Now, monomials are sometimes discussed in comparison with polynomials. The way you make a polynomial is to take at least two monomials and add or subtract them. Here are examples of some polynomials:

$x + 1$

$x^2 + x - 1$

$5x + y^2$

$1 + x - x^2 - x^3 + x^4 + \ldots + x^n$

—*Dr. Math, The Math Forum*

Defining Monomials and Polynomials

Dear Dr. Math,

I am having trouble with my homework. I need to know what monomials and polynomials are. I looked them up in the dictionary of math terms, but I didn't understand it. Can you please explain this to me in simple terms?

Aimee

Dear Aimee,

A monomial is one or more constants and/or variables that are multiplied together to make one term.

A polynomial is an algebraic expression with terms consisting of coefficients and variables, raised only to nonnegative integer powers. Here is an example:

$$2x^5 + 9x^4 - 7x^3 + 3x^2 - 5x + 1$$

This is a polynomial with six terms. In this expression, each term has a number in front and then the variable x raised to a power that is an integer. The last term (the +1), which is called the *constant* term, also has a variable, but you can't see it: it's $1x^0$—the variable is raised to the zero power.

There is no special name for a polynomial with six terms, but there are special names for some polynomials. A **binomial** has two terms—such as $3x - 5$. A **trinomial** has three terms—such as $4x^2 + x + 5$. Notice that the middle term, x, seems to have no coefficient, but it's a little like the invisible variable in the constant term: the coefficient of the middle term is 1. Whenever you don't "see" a coefficient, it is 1.

—Dr. Math, The Math Forum

Polynomials and Negative Exponents

Dear Dr. Math,

You said that polynomials have no negative exponents. This hadn't actually occurred to me before. What do you call an expression like $x^2 + 3x + y^{-4}$? Is there a reason for excluding these things from the category of polynomials, or is it a matter of definition, or what?

I've seen people give examples of polynomials with things like $x^{-\pi}$. Isn't this wrong?

I'm hoping you can explain this to me.

Sincerely,

Arturo

Dear Arturo,

For an expression to be a polynomial, the exponents must all be non-negative integer constants. I notice that some people are inconsistent: while they give negative or noninteger exponents as examples, they also give the correct standard form for a polynomial, with exponents starting at a positive integer n and decreasing to zero. It does seem like a natural extension to use negative or rational or irrational exponents, but then the expression you end up with is not a polynomial.

Why do we make this restriction? That's just the definition. I would guess that it arises from the historical origin of polynomials and the usefulness of a definition in various contexts. That is, (1) it originated before negative exponents were considered, so that exponents were thought of as merely a shorthand for multiplication, and a monomial can be considered as involving multiplication *only;* and (2) it is useful in contexts in abstract algebra where division and negative exponents need not even be defined! Moreover, many theorems involving polynomials would not apply to the broader definition. This is often what motivates a definition: It ties together a set of objects that belong in the same theorems.

As for what you call an expression that does involve negative exponents, that would be a rational expression, ultimately expressible as a ratio of polynomials.

—***Dr. Math, The Math Forum***

2 Simplifying Polynomials

In general, when you simplify a polynomial you are making the expression as short or compact as possible. As you will see as you read the following, what is "simple" to one person might not be for another.

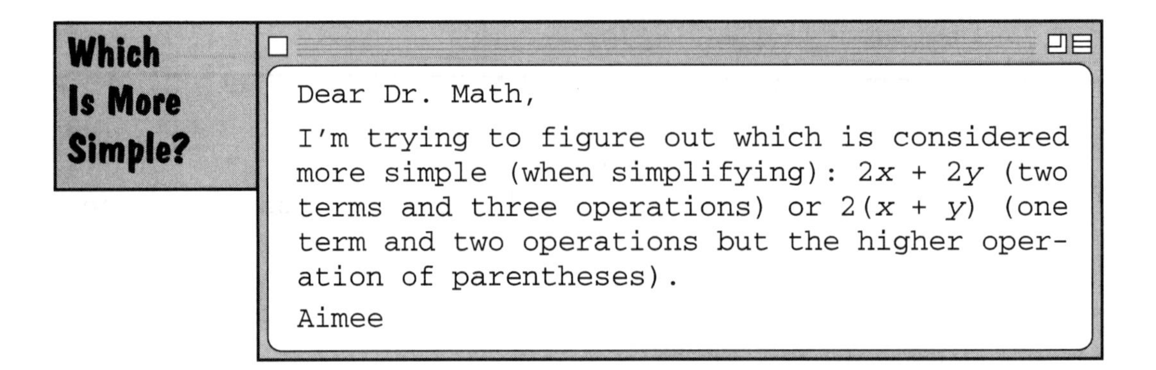

Which Is More Simple?

Dear Dr. Math,

I'm trying to figure out which is considered more simple (when simplifying): $2x + 2y$ (two terms and three operations) or $2(x + y)$ (one term and two operations but the higher operation of parentheses).

Aimee

Dear Aimee,

I often wonder the same thing; when we get questions about simplification out of context, it's hard to be sure what form is expected.

My own sense of the matter is that "simplicity" is in the eye of the beholder—and, more specifically, in the context of the problem. We simplify an expression for a reason: to make the next step easier. In some cases, the next step is just to evaluate the expression for particular values. In these cases, the form with the fewest operations may be best (unless those operations involve division by an irrational number, in which case we prefer a rationalized denominator despite a larger number of operations). At other times, if we are going to be doing something *additive,* such as looking for a polynomial, your first form is better. If we are going to be doing something *multiplicative,* the factored form is best.

In the case of your example, I would probably choose the polynomial form; everything from the order of operations rules up is built around a preference for that form and an avoidance of parentheses, so without context I naturally lean in that direction. But I'll still wonder whether it's really right . . .

—Dr. Math, The Math Forum

Evaluate and Simplify

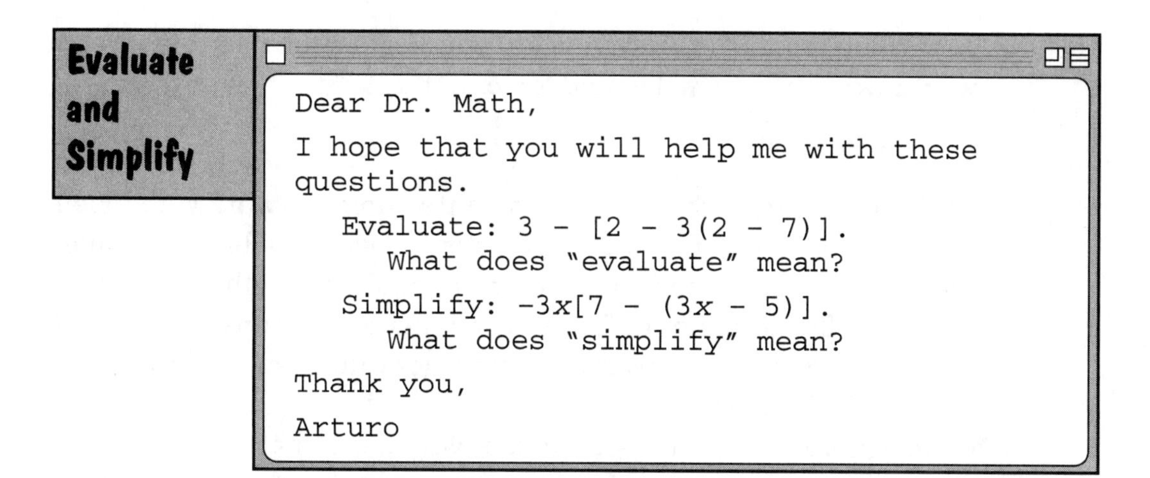

Dear Dr. Math,

I hope that you will help me with these questions.

Evaluate: 3 - [2 - 3(2 - 7)].
What does "evaluate" mean?

Simplify: -3*x*[7 - (3*x* - 5)].
What does "simplify" mean?

Thank you,

Arturo

Dear Arturo,

Evaluate means that a particular value is desired for the answer. In other words, they want a single number. **Simplify,** on the other hand, means "to make simpler," or to reduce the number of symbols used while retaining equality. So for "simplify," they don't want a single number, but instead they're looking for something like "4x" or "3x + 2"—which is as short and sweet as you can make things.

To evaluate the first problem, we start from the inside and work our way outward:

$$\begin{aligned} 3 - [2 - 3(2 - 7)] &= 3 - [2 - 3(-5)] \\ &= 3 - [2 - (-15)] \\ &= 3 - [2 + 15] \\ &= 3 - 17 \\ &= -14 \end{aligned}$$

Notice that the answer is a number (in this case, it is negative), not an expression with a variable in it.

For the second problem, here's how you could simplify:

$$\begin{aligned} -3x[7 - (3x - 5)] &= -3x[7 - 3x + 5] \\ &= -3x[12 - 3x] \\ &= 9x^2 - 36x \end{aligned}$$

Now, although this is probably the "simplest" form, it isn't the way I would want to write it. I would factor out a "9x" from each term.

$$= 9x(x - 4)$$

I'd rather say it this way because it's easier to work with when you put it together with other expressions. If you had this expression on top of a fraction, you could see more easily if anything could be canceled. But in general, "simple" means fewer parentheses and groupings, as well as fewer operations (addition, subtraction, multiplication, division) used.

In general, it doesn't really matter whether somebody says "simplify" or "evaluate." They just want you to make the expression as short or compact as possible. The technical difference is that unlike "evaluate," where your answer is a single number, "simplify" usually means ending up with an expression with variables. You (usually) can't get a single number as the answer.

—*Dr. Math, The Math Forum*

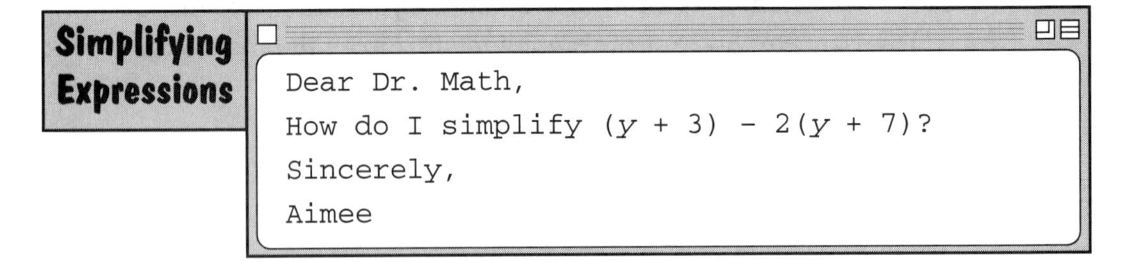
Simplifying Expressions

Dear Dr. Math,

How do I simplify $(y + 3) - 2(y + 7)$?

Sincerely,

Aimee

Dear Aimee,

Let me show you how to do a similar problem and then perhaps you will have an easier time with the one you sent us.

$$(y + 6) - 3(y + 4)$$

The pieces inside the parentheses can't be simplified yet, so I will keep them together and multiply $3(y + 4)$ out to get $3y + 12$. Now my problem looks like:

$$(y + 6) - (3y + 12)$$

Now we can subtract, but we have to *distribute* the minus sign like this:

$$(y + 6) - 3y - 12$$

Now I can add like terms ($-3y$ to y and 6 to -12) to get:

$$-2y - 6$$

which is the simplest form of the expression. Can you use these same steps to solve your problem?

—Dr. Math, The Math Forum

Examples of Simplifying Polynomials

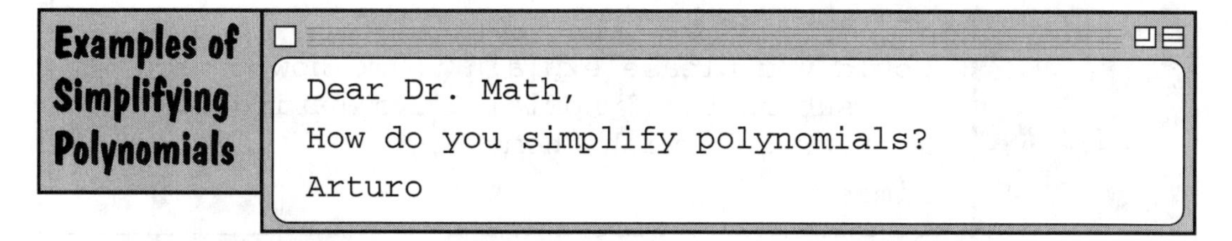

Dear Dr. Math,

How do you simplify polynomials?

Arturo

Dear Arturo,

The word "simplify" is open to interpretation, but I'll give you a few examples:

Here are three ways of writing the exact same polynomial:

Version one: $3x^3 + 7x^2 - 2x - 8$

Version two: $(x + 2) \cdot (3x + 4) \cdot (x - 1)$

Version three: $((3x + 7) \cdot x - 2) \cdot x - 8$

Version one is in so-called standard form and clearly shows the coefficients of each power of the variable x.

Version two is factored, which makes it easy to see which x values make that product of three things equal to zero. For instance, if $x = 1$, then the last factor is zero, which makes the whole thing zero. If $x = -2$, then the first factor is zero. What x value makes the middle factor zero?

Version three may *look* extra complicated, but it is just right for computer programs. It is a very speed-efficient way to evaluate the polynomial expression, given an x value.

—Dr. Math, The Math Forum

Adding and Subtracting Polynomials

Just as we can add and subtract numbers, we can also use these operations with polynomials.

One of the keys to success with this topic is lining things up properly. Remember in arithmetic when it was important to line up the decimal points if you were adding several numbers? Well, this is similar. Keep reading to find some ideas that will help you.

Dear Dr. Math,

Could you please explain to me how to add and subtract polynomials? I'm having so much trouble with them!

Aimee

Dear Aimee,

I happen to have some examples right here.

Add the following:

$$(4x - 2x^2 - 7xy) + (2x^2 + 5xy)$$

When you are adding (or subtracting) polynomials, you must find the variables and exponents that match. For example, in the problem above, $-2x^2$ and $2x^2$ have the same variable and exponent. So do $-7xy$ and $5xy$. Unless the variables and the exponents both match exactly, you can't add (or subtract) them. Once you find that something matches (like $-2x^2$ and $2x^2$), then you can add or subtract the coefficients. Here's how this works using our example:

$4x$	There is nothing to add with it, so we leave it alone, but include it in the answer.
$-2x^2 + 2x^2 = 0x^2$	So we won't have an x^2 term in the answer.
$-7xy + 5xy = -2xy$	

The answer is $4x - 2xy$.

Here's another addition example:

$$(4 - 5x^2 + 7x^3) + (4x^3 + 5x^2 + 5x^4)$$

We usually start with the highest power of the variable and work our way down.

$5x^4$	Nothing to add to it, so we leave it be.
$7x^3 + 4x^3 = 11x^3$	
$-5x^2 + 5x^2 = 0x^2$	So there will be no x^2 term in our answer.
4	Nothing to add to it, so we leave it alone.

The answer is $5x^4 + 11x^3 + 4$.

Subtraction is very similar to addition. You can think of it in one of two ways. You can think of it as subtracting the coefficients (instead of adding them), or you can think of it as adding the negative of the coefficients.

Here are a couple of subtraction problems:

$$(8x^3 + x^2 - 7x - 11) - (5x^3 + 3x^2 - 3x + 8)$$

When I teach this to my students, I tell them to go through and distribute the negative sign to the second group. This is because sometimes when you subtract, there isn't a like term in the first group. However, you still need to make the sign in front of the number you're trying to subtract the opposite of what you are given. By distributing the negative sign, you make sure to properly take care of these terms.

So I would do this first:

$$(8x^3 + x^2 - 7x - 11) - 5x^3 - 3x^2 + 3x - 8$$

Now we have a situation like the one we had before with addition:

$$8x^3 - 5x^3 = 3x^3$$

$$x^2 - 3x^2 = -2x^2$$

$-7x + 3x = -4x$

$-11 - 8 = -19$

So the answer is: $3x^3 - 2x^2 - 4x - 19$.

Here's another subtraction example:

$(4a^2 - 6a) - (2a^2 + 5a - 3)$

Distributing the negative sign, I get:

$(4a^2 - 6a) - 2a^2 - 5a + 3$

Now I combine like terms:

$4a^2 - 2a^2 = 2a^2$

$-6a - 5a = -11a$

3 (this stays as 3 since there is nothing to add to it in the first set)

So the answer is: $2a^2 - 11a + 3$.

—Dr. Math, The Math Forum

Multiplying Polynomials

To multiply two polynomials, simply multiply each element of the first by each element of the second. The coefficients (the numbers before the variables) are multiplied, and the exponents are added.

For example:

$$(2x^3 + x^2 + 1)(x + 3)$$

$$= 2x^3(x) + 2x^3(3) + x^2(x) + x^2(3) + 1(x) + 1(3)$$

$$= 2x^4 + 6x^3 + x^3 + 3x^2 + x + 3$$

It is usually good to group all the terms with the same exponent together in descending order, so in the end you would have this:

$$2x^4 + 7x^3 + 3x^2 + x + 3$$

Distributive Property

Dear Dr. Math,

How do you actually work out this kind of problem?

Directions: Simplify each expression.

$12(c+3d+4e)+2(2c+d+6e)$

It's really confusing when I have those big math problems. This one is kind of hard too:

$3(9ab + 8ab - 7ab)$

Arturo

Dear Arturo,

I can see how it can be confusing. There's a lot to do. Let's try to break it down into simpler tasks—then it might seem less confusing.

The first thing I would do is put some space between the different parts of the expression:

$$12(c + 3d + 4e) + 2(2c + d + 6e)$$

It seems less cluttered to me when I write it like this, and that helps unclutter my thinking.

It looks as if we have some multiplying and some adding to do. You have to do multiplication in two parts of the problem and then add those results together. Let's look at just one part of the problem where you have to do multiplication:

$$12(c + 3d + 4e)$$

We can't add the items in the parentheses to one another because the variables are different. If there were, say, two "c" terms, you could add them, but there aren't in this case. Here, the only thing there is to do is multiply the 12 by each of the items inside the parentheses (applying the distributive property over addition).

After multiplying 12 by each of the items in the parentheses, we get:

$$12c + 36d + 48e$$

We can't add these items together, either. So this part is as simple as it can get. Now it's time to look at the other part of the problem:

$$2(2c + d + 6e)$$

It looks as if we need to do the same thing to it, for the same reasons, so:

$$4c + 2d + 12e$$

Do you see how I did that? Looking back at the original problem, it shows that these two parts of the problem need to be added together. Let's put them side by side again, with an addition sign between them, to make it clear:

$$12c + 36d + 48e \quad + \quad 4c + 2d + 12e$$

Because they are all plus signs, it doesn't matter if there are parentheses, and we can add them in any order we like, as long as it makes sense. In this case, the thing to do is add the terms that share the same variable, also called "like terms." So,

$$12c + 4c = 16c$$

$$36d + 2d = 38d$$

$$48e + 12e = 60e$$

Giving us:

$$16c + 38d + 60e$$

Can you simplify this expression any more?

Try the second problem that you wrote above, doing it the same way as I did the last part for the first problem. You will probably see what to do.

—Dr. Math, The Math Forum

Expanding Polynomials

Dear Dr. Math,

Help! I have a math exam on Friday, and I don't understand how to expand three binomials multiplied together.

My example question is: $(s + 4)(2s - 1)(s - 2)$. I have the answer sheet, but I can't get the answer. I am also having problems with this question: $(z^2 - 2z + 3)(2x^3 + z - 1)$.

I have tried the "foil" method for both of them, but I got both wrong.

Sincerely,

Aimee

Dear Aimee,

The FOIL method (First, Outside, Inside, Last) only applies to a product of two binomials.

$$(s + 4)(2s - 1) = \underset{\text{First}}{(s)(2s)} + \underset{\text{Outside}}{(s)(-1)} + \underset{\text{Inside}}{(4)(2s)} + \underset{\text{Last}}{(4)(-1)}$$

FOIL is a particular example of using the distributive property more than once. I'll show you how the distributive property gives you FOIL.

First, keep the second binomial intact, but break up the first binomial using the distributive property:

$$(a + b) \quad c \quad = a \quad c \quad + b \quad c$$
$$(s + 4)(2s - 1) = s(2s - 1) + 4(2s - 1)$$

Next, in each of the two terms that you got from the first application of the distributive property, use the property again:

$$a(b + c) = a \; b \; + a \; c$$
$$s(2s - 1) = s(2s) + s(-1)$$
$$4(2s - 1) = 4(2s) + 4(-1)$$

Now you have four terms, and they are the same as those that we would have gotten using FOIL. Does this give you ideas? The distributive property is much more flexible than FOIL; we can apply it to trinomials or more. I'll show you:

$$(a + b + c) \quad d \quad = a \quad d \quad + \quad b \quad d \quad + c \quad d$$
$$(z^2 - 2z + 3)(2z^2 + z - 1) = z^2(2z^2 + z - 1) + (-2z)(2z^2 + z - 1) + 3(2z^2 + z - 1)$$

$$a\,(b + c + d) = a \; b \; + a\; c \; + a \; d$$
$$z^2(2z^2 + z - 1) = z^2(2z^2) + z^2(z) + z^2(-1)$$

I'll let you expand the other two terms. You will end up with nine terms. You can "collect terms" by grouping all those with the same power of z and using the distributive property in the other direction to "pull out" the power of z. For instance, you will have terms:

$$\begin{aligned} z^2(z) + (-2z)(2z^2) &= z^3 - 4z^3 \\ &= 1z^3 - 4z^3 \\ &= (1 - 4)z^3 \\ &= -3z^3 \end{aligned}$$

Do you see how to expand products of trinomials now? You can use the same method to expand products of any two polynomials. But you can also expand products of more than two polynomials, as in your first example:

$$(s + 4)(2s - 1)(s - 2)$$

Just expand the product of the first two factors; it will become:

$$\begin{aligned} (s + 4)(2s - 1) &= 2s^2 - s + 8s - 4 \\ &= 2s^2 + 7s - 4 \end{aligned}$$

Replace the first two factors in your example with this expansion:

$$(2s^2 + 7s - 4)(s - 2)$$

Expand this product of polynomials, and you're done.

The hardest part of this is making sure you don't lose a term. Just keep it organized as I did, and you won't have much trouble. When

I regroup terms according to the power of the variable, I cross out a term as I use it, so I can tell when I have used them all.

—***Dr. Math, The Math Forum***

Explaining Algebra Concepts

Dear Dr. Math,

I have been trying and trying to help my friend, but he just doesn't get it. I started off simple, with $x + 22 = 45$, and he got that, but when we did anything else he got lost. For example, he just doesn't understand $4y(5y - 3) + 3y(y + 4)$.

I have tried teaching him FOIL, but he can't do it. If you have any advice, please let me know.

Sincerely,

Arturo

Dear Arturo,

One thing I like to do is to not only start simple but keep it simple. That means avoiding too many new ideas and just explaining everything in terms of basic math. For instance, I avoid FOIL, because it just looks like one more recipe to memorize, when it's really just a way to keep track of what you have to multiply when you use the distributive rule. Also, it doesn't apply to multiplying trinomials, so it can get in the way of a proper understanding of algebra. What I like to do is to model multiplication of polynomials after multiplication of numbers, so it looks familiar.

For example, for $4y(5y - 3)$, I would write:

$$\begin{array}{r} 5y - 3 \\ \times \quad 4y \\ \hline 20y^2 - 12y \end{array}$$

and for $(a - b)(c + d)$ I would write:

$$\begin{array}{r} c + d \\ \times\ a - b \\ \hline -bc - bd \\ ac + ad \qquad\quad \\ \hline ac + ad - bc - bd \\ (F \quad O \quad I \quad L) \end{array}$$

This is most useful when you can group like terms as you go (which is really exactly what you do when you multiply numbers):

$$\begin{array}{rrr} & 3x & +2 \\ \times & 2x & -3 \\ \hline & -9x & -6 \\ 6x^2 & +4x & \\ \hline 6x^2 & -5x & -6 \\ (F & O+I & L) \end{array}$$

Do you see how this helps to organize complex multiplications without adding new ideas to worry about? It lets us concentrate on the important idea, which is that multiplication distributes over addition, meaning that each term inside a set of parentheses has to be multiplied by what's outside the parentheses. Once you understand distribution thoroughly, it should all fall into place, especially when you have a technique to keep the work organized. Until you have that understanding, no amount of technique will help. Since it sounds like this is a major area of difficulty, you should try to find out whether your friend needs more work on distribution or is just overwhelmed by complex problems and needs techniques like this to help keep things simple.

A similar thing happens in solving equations. Simple equations can be easy, but when there are too many things happening in an equation, they can be distracting. Your friend may need to learn how to focus on one part of an equation at a time. I like to talk about this as "peeling an onion one layer at a time" or as "shucking corn." Have you ever noticed that when you take the husks off an ear of corn, if you try to do it one leaf at a time, you have to look around for the outermost leaf and pull it off, so no other leaf gets in the way? When you solve a complicated equation, you have to look for the "outermost" part of the expression and pull that off, ignoring the rest of it.

Here's what I mean. If we have to solve:

$$6(3(4x - 2) - 4) + 3 = 0$$

the parentheses protect the inner part, so before we get to them we have to remove the 3 (by adding –3 to both sides) and then the 6 (by dividing both sides by 3):

$$6(3(4x - 2) - 4) = -3$$

$$3(4x - 2) - 4 = \frac{-3}{6}$$

Now we can work on the next layer.

(Of course, you can also work on this sort of problem from the inside out, simplifying the expression so there aren't so many layers, before you start peeling off what's left.)

—Dr. Math, The Math Forum

Patterns in Polynomials

As stated earlier, a polynomial that contains two terms is called a *binomial* and one that contains three terms is called a *trinomial*. If we square a binomial (multiply it by itself), we get a trinomial. We can write this in general form like this:

$$(x + y)^2 = x^2 + 2xy + y^2$$

Here's a picture of the idea so that you can see the geometric thought behind the algebraic notation:

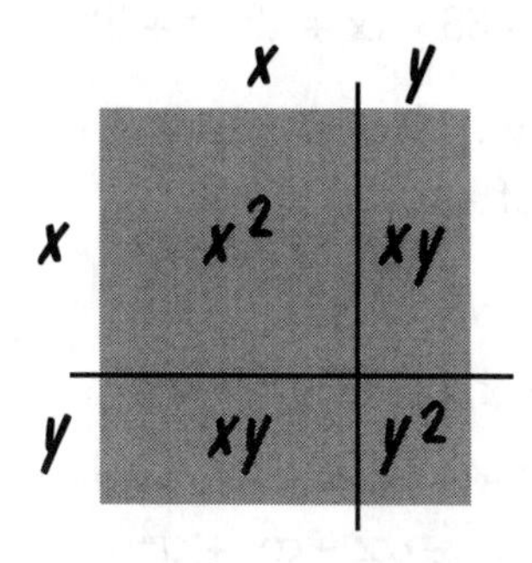

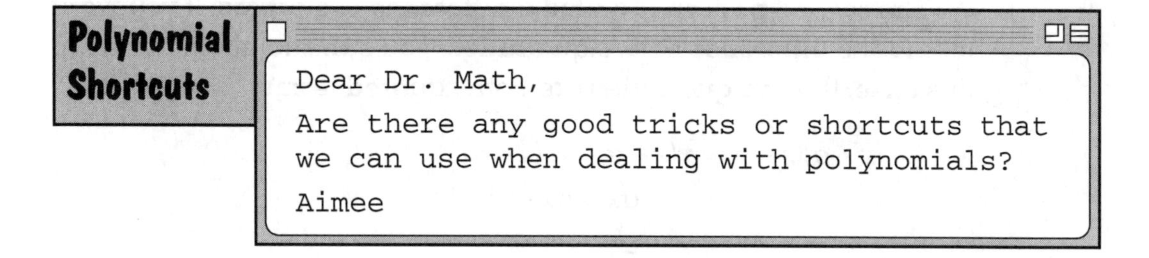
Polynomial Shortcuts

Dear Dr. Math,

Are there any good tricks or shortcuts that we can use when dealing with polynomials?

Aimee

Dear Aimee,

There are two patterns that come up so often that they're worth memorizing. The first is the *square of a binomial:*

$$\begin{aligned}(x + a)^2 &= (x + a)(x + a)\\ &= x(x + a) + a(x + a)\\ &= x^2 + ax + ax + a^2\\ &= x^2 + 2ax + a^2\end{aligned}$$

Whenever you see a **quadratic** polynomial (a polynomial with one variable, whose highest exponent is 2) in which the constant term is a perfect square, a little bell should go off in your head, telling you to consider whether this might be a binomial squared. Find the square root of the constant term, double it, and check that against the coefficient of the linear term. Sometimes it will work,

$$x^2 + 12x + 36 = (x + 6)^2 \qquad 2 \cdot \sqrt{36} = 12 \qquad \text{Yes!}$$

and sometimes it won't,

$$x^2 + 15x + 36 = (x + 12)(x + 3) \qquad 2 \cdot \sqrt{36} = 12 \qquad \text{Doh!}$$

but it's always worth checking. A variation on this has a slightly different pattern of signs:

$$\begin{aligned}(x - a)^2 &= (x - a)(x - a)\\ &= x(x - a) - a(x - a)\\ &= (x^2 - ax) - (ax - a^2)\\ &= x^2 - ax - ax + a^2\\ &= x^2 - 2ax + a^2\end{aligned}$$

The second pattern is called a **difference of squares.** If you want to find the difference of two quantities, you subtract them, right? In this case, the two quantities are both squared terms:

$$\begin{aligned}(x + a)(x - a) &= x(x - a) + a(x - a)\\ &= x^2 - ax + ax - a^2\\ &= x^2 - a^2\end{aligned}$$

Again, the tip-off is a constant term that happens to be a perfect square. The confirmation is a missing linear term.

These patterns have interesting visual representations:

$$(x + a)^2 = x^2 + 2ax + a^2$$

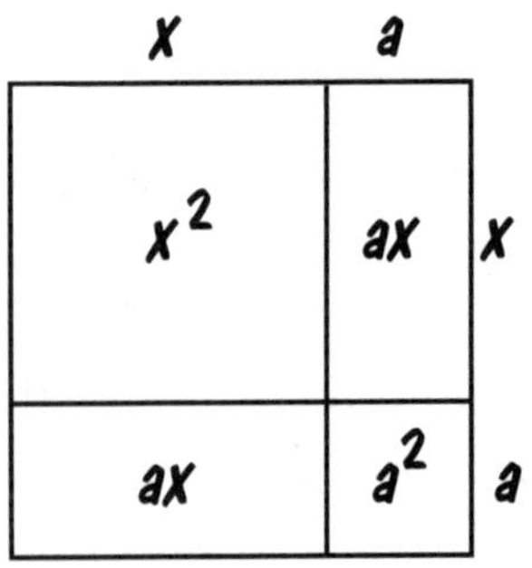

	x	a	
	x^2	ax	x
	ax	a^2	a

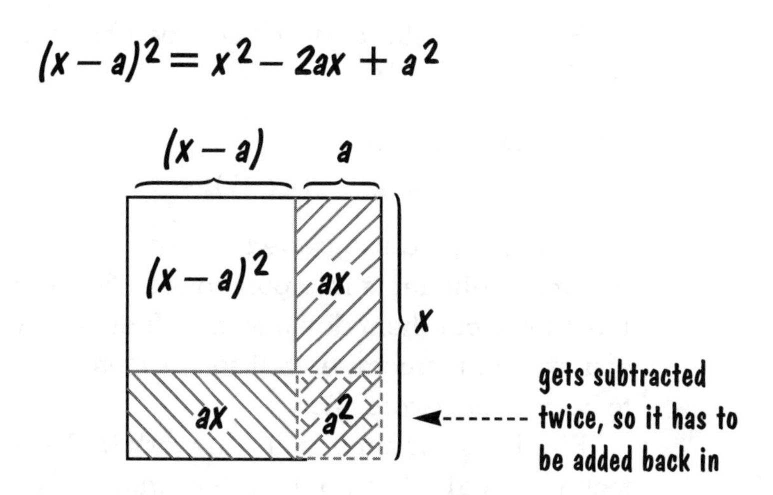

$$(x + a)(x - a) = x^2 - a^2$$

x + a
x
x – a
(x – a)²
a(x–a)
a
a(x – a)
a²

One other thing to keep in mind is that sometimes these come disguised as higher-degree polynomials. For example,

$$x^6 - 8x^3 + 16$$

In this case, we can think of x^6 as $(x^3)^2$, which makes our "variable" x^3 instead of just x:

$$\begin{aligned} x^6 - 8x^3 + 16 &= (x^3)^2 - 8(x^3) + 16 \\ &= (x^3 - 4)^2 \end{aligned}$$

Sometimes the patterns can be obscured a little by a leading coefficient:

$$16x^2 - 49 = (4x)^2 - 7^2$$
$$= (4x + 7)(4x - 7)$$

So don't get discouraged just because you see big exponents or leading coefficients in a polynomial! Sometimes you can use tricks like this to cut them down to size. Just keep your eye out for perfect squares, and remember that sometimes your "variable" will turn out to be a whole expression.

Finally, sometimes you can work in the other direction, using a technique called *completing the square*. To do this, instead of looking for a perfect square in the constant term, you just look for an even coefficient in the linear term. For example,

$$y = x^2 + 10x - 4$$

The next step is to use a little imagination. If this *were* the square of a binomial with a general form of $x^2 + 2ax + a^2$, it would look like:

$$x^2 + 10x - 25$$

because if 10 is equal to $2a$, then 5 must be equal to a, and 25 must be equal to a^2. Does that make sense? So what happens if we just add 25 to both sides of the equation?

$$y + 25 = x^2 + 10x + 25 - 4$$
$$= (x + 5)^2 - 4$$

Now we can subtract the same 25 to get back to where we were:

$$y + 25 - 25 = (x + 5)^2 - 4 - 25$$
$$y = (x + 5)^2 - 29$$

Why would we want to do this? Well, it turns out that

$$y = (x - h)^2 + k$$

is the equation of a **parabola** whose **vertex** is at the point (h, k). So this tells us that we have a parabola whose vertex is at $(-5, -29)$, something that isn't at all obvious from the original form of the equation.

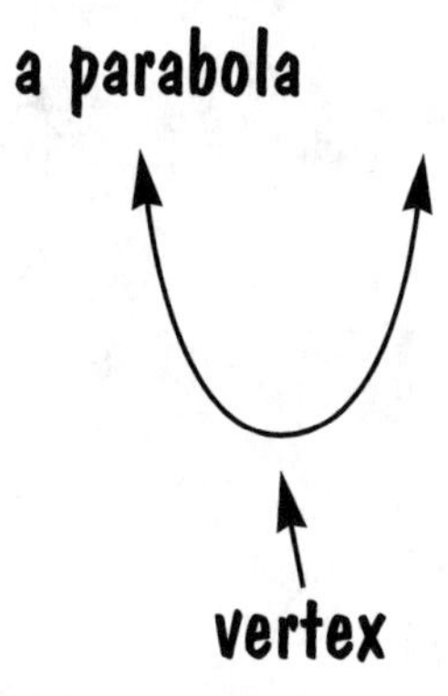

I know that this is a lot to take in at once! So don't feel too bad if you can't just sit down and start applying these techniques right away. All I'm trying to do here is give you a preview of things that you'll be seeing later on, so they won't seem so overwhelming then.

Why is it important to be able to recognize and apply these patterns? Well, remember back when you were learning to simplify fractions? The easiest way to do it was to break the numerator and denominator into factors, and cancel out the **common factors** to get something simpler:

$$\frac{24}{52} = \frac{\overset{\downarrow}{2} \cdot \overset{\downarrow}{2} \cdot 2 \cdot 3}{\underset{\uparrow}{2} \cdot \underset{\uparrow}{2} \cdot 13} = \frac{6}{13}$$

Well, you'll have lots of opportunities to do similar things with polynomials, for example,

$$\frac{x^2 - 9}{x^2 + 6x + 9} = \frac{\overset{\downarrow}{(x+3)}(x-3)}{\underset{\uparrow}{(x+3)}(x+3)} = \frac{(x-3)}{(x+3)}$$

—*Dr. Math, The Math Forum*

6 Dividing Polynomials

It's possible to divide a polynomial by another polynomial of equal or lesser degree. For example, you can divide a cubic polynomial (degree 3) by a quadratic one (degree 2). Sometimes division looks like what you do when you simplify fractions. Other times it looks like long division—except that instead of lining up equivalent place values, you have to line up terms with equivalent exponents.

Dividing Polynomials by Monomials

Dear Dr. Math,

$$\frac{12ax + 16x}{4x} = ?$$

I just don't understand how to do this.

Arturo

Dear Arturo,

This is a good place to remember your fractions. We're going to treat this expression as if it were a simple fraction: Factor the numerator and denominator, cancel the things that are the same in both, and see what we end up with.

We start with

$$12ax + 16x$$

The first thing I notice is that both expressions have an x in them, so we can pull it out. We do that by dividing both terms by x and putting them in a pair of parentheses, which we indicate is then multiplied by x, like this:

$$x(12a + 16)$$

Next, I look at 12 and 16 (especially since I already know I have to divide by 4x) and realize they're both divisible by 4. So I can pull the 4 out the same way I pulled out the x, and I get:

$$4x(3a + 4)$$

Now the problem is easy, because I have 4x times something divided by 4x, and the 4x's cancel each other out (because anything divided by itself is 1). So I end up with:

$$\frac{4x(3a + 4)}{4x} = 3a + 4$$

Does this help? Whenever you get a problem like this, the first thing to try is to see if the part you're dividing by (the 4x, in this case) is a factor of each expression in the numerator.

You'd do the same thing if you had

$$\frac{(3 \cdot 2 \cdot 5) + (3 \cdot 7)}{3}$$

You'd know to look for "3" as a common factor in each expression.

Extra credit: Can you find the answer to the problem I just posed?

—Dr. Math, The Math Forum

Polynomial Long Division

Dear Dr. Math,

I am trying to figure out how to solve this problem:

$$\frac{x^3 + 3x^2 - 4x + 3}{x + 1}$$

If you can walk me through this and give me a formula that I can always apply to get to the right answer, I'd really appreciate it.

Sincerely,

Aimee

Dear Aimee,

Here is how to proceed, using long division:

$$x+1 \overline{)\, x^3 + 3x^2 - 4x + 3}$$

Divide the leading term "x" of the divisor into the leading term "x^3" of the dividend. The quotient is x^2. Put that above the $3x^2$ term of the dividend, and above the line. Multiply that x^2 times the divisor up to this point, putting the result below the dividend, and subtract. Bring down the next term from the dividend into the current remainder. Your work should look like this:

$$\begin{array}{r} x^2 \\ x+1 \overline{)\, x^3 + 3x^2 - 4x + 3} \\ \underline{x^3 + x^2} \\ 2x^2 - 4x \end{array}$$

Now divide x into $2x^2$ to get the next term of the quotient. It is $2x$. Put it above the $-4x$ term of the dividend and above the line. Multiply that $2x$ times the divisor, putting the result below the current remainder, and subtract. Bring down the next term from the dividend into the current remainder. Now you should have:

$$
\begin{array}{rl}
 & x^2 + 2x \\
x+1 & \overline{)\,x^3 + 3x^2 - 4x + 3} \\
 & \underline{x^3 + x^2} \\
 & \quad 2x^2 - 4x \\
 & \quad \underline{2x^2 + 2x} \\
 & \qquad\quad -6x + 3
\end{array}
$$

Now divide x into –6x to get the next term of the quotient. It is –6. Put it above the +3 term of the dividend and above the line. Multiply that –6 times the divisor, putting the result below the current remainder, and subtract. Now you should have:

$$
\begin{array}{rl}
 & x^2 + 2x - 6 \\
x+1 & \overline{)\,x^3 + 3x^2 - 4x + 3} \\
 & \underline{x^3 + x^2} \\
 & \quad 2x^2 - 4x \\
 & \quad \underline{2x^2 + 2x} \\
 & \qquad\quad -6x + 3 \\
 & \qquad\quad \underline{-6x - 6} \\
 & \qquad\qquad\qquad 9
\end{array}
$$

You are done, because you cannot divide $x + 1$ into 9. The quotient is the top line, $x^2 + 2x - 6$, and the remainder is the bottom line, 9. This means $(x^3 + 3x^2 - 4x + 3)/(x + 1)$ is $x^2 + 2x - 6$ with remainder 9, or in other words,

$$\frac{x^3 + 3x^2 - 4x + 3}{x + 1} = x^2 + 2x - 6 + \frac{9}{x + 1}$$

Notice that, reading down the columns, every term has the same exponent of x. This is the best way to arrange your work. Keeping your columns straight will help you keep from getting confused.

—Dr. Math, The Math Forum

Resources on the Web

Learn more about polynomials at these Math Forum sites:

Algebra Problem of the Week: Triangular Number Identities

mathforum.org/algpow/solutions/solution.ehtml?puzzle=55

Prove two famous theorems of ancient number theory: the sum of any two consecutive triangular numbers is a square number, and the product of 8 and any triangular number, increased by 1, is a square number.

Algebra Problem of the Week: What's What?

mathforum.org/algpow/solutions/solution.ehtml?puzzle=68

A poetical puzzle to ponder about: "What is WHAT?"

Algebra Problem of the Week: My Bag of Marbles

mathforum.org/algpow/solutions/solution.ehtml?puzzle=104

Find the probability of drawing a blue marble after additional marbles have been placed in a bag.

Graphing Polynomial Functions

mathforum.org/alejandre/polynomial.graph.html

Step-by-step directions to create graphs of polynomial functions using a ClarisWorks spreadsheet file.

Factoring

One big reason for learning how to factor numbers is to make it easy to reduce fractions. That's true for factoring polynomials, too (as you saw in part 3). Another reason is that factoring can make some information obvious that is hidden in other forms. For example, from

$$(x - 2)(x - 3) = 0$$

it's obvious that the equation can be true only when $x = 2$ or $x = 3$, something that isn't clear at all in the form

$$x^2 - 5x + 6 = 0$$

In the end, the power of factoring can be summed up in one word: "uniqueness." There is only one way to break a number into prime factors, so when you've found it, you know (or can easily find out) just about everything about that number. Similarly, there is only one way to factor a polynomial into binomials, and when you've done that, you've got that polynomial right where you want it.

In this part, Dr. Math explains:

- prime, composite, and square numbers
- factoring monomials
- factoring polynomials
- factoring the difference of two squares

1 Prime, Composite, and Square Numbers

Prime factoring is taking a composite number and splitting it up into the smaller factors that it's made up of, until you can't split it up any more. Recognizing whether a number is prime or composite just helps you know if you should take the time to factor it. If a number is prime, you needn't bother. You won't be able to "break it up" into any other whole numbers. It's actually handy to take a little time to memorize the first 10 prime numbers: 2, 3, 5, 7, 11, 13, 17, 19, 23, and 29.

A square number is a special type of composite, and squares turn out to be very useful in factoring because they can be easily recognized. If you get in the habit of recognizing square numbers, such as 4, 9, 16, 25, 36, and so on, they can help you with factoring.

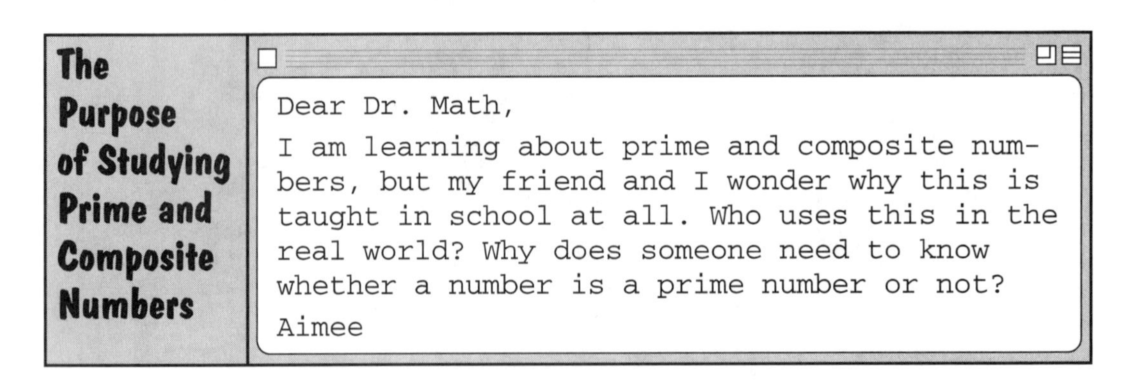

The Purpose of Studying Prime and Composite Numbers

Dear Dr. Math,

I am learning about prime and composite numbers, but my friend and I wonder why this is taught in school at all. Who uses this in the real world? Why does someone need to know whether a number is a prime number or not?

Aimee

Dear Aimee,

Every time someone sends a credit card number over the Internet, it gets encrypted by a browser, and the encryption algorithm is based on the theory of prime numbers. At some point, electronic money may become as common as paper money, and *that* will also be based on the theory of prime numbers. And what's used more in the real world than money?

The importance of prime numbers in algebra is that any integer can be decomposed (broken down) into a product of primes. For example, if you want to know how many different pairs of numbers can be multiplied to get 360, you can start by trying to write them down,

$1 \cdot 360$

$2 \cdot 180$

$3 \cdot 120$

$4 \cdot 90$

$5 \cdot 72$

$6 \cdot 60$

checking every single number up to 180, and hoping that you don't miss any; or you can decompose 360 into its prime factors,

$$360 = 2 \cdot 2 \cdot 2 \cdot 3 \cdot 3 \cdot 5$$

with the assurance that every factor of 360 will be a product of a subset of these prime factors. Let's look at that for a minute. If you multiply $2 \cdot 2$, you get 4. In the factor pairs that we listed above, 4 is paired with 90. 90 is $2 \cdot 3 \cdot 3 \cdot 5$. Well, 2, 2, 2, 3, 3, and 5 are the prime factors of 360. If you are systematic, you can check the prime factors in this way to find all of the factor pairs.

This kind of analysis is extremely convenient when working with fractions (since prime factorization tells you which common denominators are available for any two fractions), when factoring polynomials, when doing just about anything where integers are involved, really.

Think of it this way. You don't need to learn to multiply, since you can always use repeated addition to solve any multiplication problem, right? If you want to know what 398 times 4,612 is, you can just start adding:

```
 398  (1)
 398  (2)
----
 796
 398  (3)
----
1194
 398  (4)
----
and so on
```

Knowing how to multiply saves you time. That's all it does—but that's a lot!

Mostly, prime numbers are good for quickly transforming a situation with zillions of possible outcomes into an equivalent situation with only a handful of possible outcomes.

—Dr. Math, The Math Forum

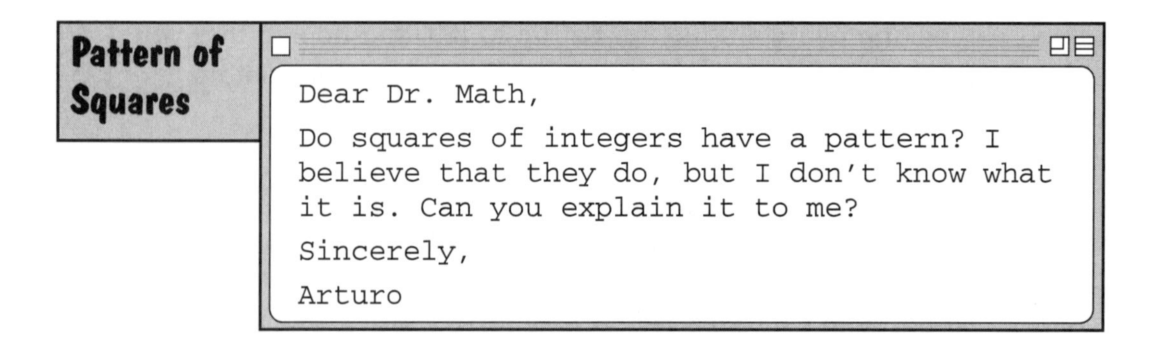

Pattern of Squares

Dear Dr. Math,

Do squares of integers have a pattern? I believe that they do, but I don't know what it is. Can you explain it to me?

Sincerely,

Arturo

Dear Arturo,

There is a very definite pattern to the squares of integers. I'll show you how to find it.

Pick a number, any number. I picked 7. Now say that you have that number of dots. I have 7 dots, and there is not much I can do with them. I could put them in a line:

■ ■ ■ ■ ■ ■ ■

or arrange them in two unequal rows:

■ ■ ■

■ ■ ■ ■

but it's not too pretty.

But say I had 9 dots. Then I could arrange them in a perfect square:

■ ■ ■

■ ■ ■

■ ■ ■

If you are getting ahead of me, you'll notice that 9 is 3^2.

In fact, if the number of dots is the square of a positive integer, you can always arrange them in a square.

For example, $5^2 = 25$, and we can make a 5-by-5 square out of the 25 dots.

Now that you've got that, let's look at successive squares. Our first square is 1 by 1, or just 1 dot.

■

Our second square is 2 by 2, or 4 dots:

■ ■

■ ■

Notice the original dot in the upper right hand corner. Our third square is 3 by 3, with 9 dots.

■ ■ ■

■ ■ ■

■ ■ ■

Our fourth square is 4 by 4, with $16 = 4^2$ dots.

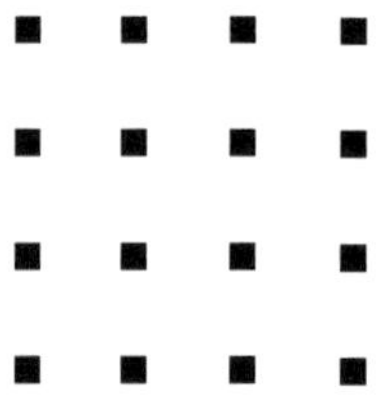

■ ■ ■ ■

■ ■ ■ ■

■ ■ ■ ■

■ ■ ■ ■

Are you starting to see a pattern? Notice that with each successive square, we have to add a new row and a new column. The number of dots we have to add each time is odd. Do you see why?

Okay, now we're ready to go back and look for the pattern. Our first square had 1 dot in it.

To make our second square, we had to add 3 dots. So our second square (which had 2^2 dots in it) has $1 + 3$ dots.

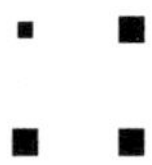

▪ ■

■ ■

Our third square (with $3^2 = 9$ dots) we got by adding 5 dots to the 2-by-2 square. Thus the total number is $1 + 3 + 5$.

▪ ▪ ■

▪ ▪ ■

■ ■ ■

Likewise, the number of dots in our fourth square is $1 + 3 + 5 + 7$.

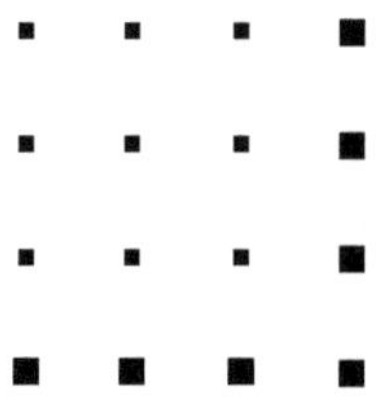

▪ ▪ ▪ ■

▪ ▪ ▪ ■

▪ ▪ ▪ ■

■ ■ ■ ■

Do you see that pattern? What happens in the 5-by-5 square?

—Dr. Math, The Math Forum

Can you answer Dr. Math's question?
Yes, I see the pattern. The next number after 1+3+5+7 would be +9, so you would have 4 dots added to the bottom, 4 dots added to the side, and one dot added to the corner.

PRIME NUMBERS

What is a prime number?

A prime number is a positive integer that has exactly two positive integer factors, 1 and itself. For example, if we list the factors of 28, we have 1, 2, 4, 7, 14, and 28. That's six factors. If we list the factors of 29, we only have 1 and 29. That's two. So we say that 29 is a prime number, but 28 isn't.

Another way of saying this is that a prime number is a whole number that is not the product of two smaller numbers.

Note that the definition of a prime number doesn't allow 1 to be a prime number: 1 only has one factor; namely, 1. Prime numbers have *exactly* two factors, not "at most two" or anything like that.

When a number has more than two factors it is called a composite number.

Here are the first few prime numbers:

2, 3, 5, 7, 11, 13, 17, 19, 23, 29, 31, 37, 41, 43, 47, 53, 59, 61, 67, 71,

73, 79, 83, 89, 97, 101, 103, 107, 109, 113, 127, 131, 137, 139, 149,

151, 157, 163, 167, 173, 179, 181, 191, 193, 197, 199.

What's a quick method of finding prime numbers?

Eratosthenes (275–194 B.C., Greece) devised a "sieve" to discover prime numbers. A sieve is like a strainer that you use to drain spaghetti when it is done cooking. The water drains out, leaving your spaghetti behind. Eratosthenes's sieve drains out composite numbers and leaves prime numbers behind.

To use the sieve of Eratosthenes to find the prime numbers up to 100, first make a chart of the first one hundred whole numbers (1–100). Then follow these steps:

1. Cross out 1, because it is not prime.
2. Circle 2, because it is the smallest positive even prime. Now cross out every multiple of 2; in other words, cross out every second number.
3. Circle 3, the next prime. Then cross out all of the multiples of 3; in other words, every third number. Some, like 6, may have already been crossed out because they are multiples of 2.
4. Circle the next open number, 5. Now cross out all of the multiples of 5, or every fifth number.

Continue doing this until all the numbers through 100 have been either circled or crossed out. You have just circled all the prime numbers from 1 to 100!

What's the largest known prime?

There is no largest prime number, but the effort to find ever-larger primes is ongoing. The easiest way to generate such a list is to use a computer program that tests each number to see if it can be factored by any number up to and including that number's square root. You can stop at the square root because all other combinations of factors would include one number larger than the square root, and one smaller. Once you have gone through the smaller numbers, it would be redundant to check the larger ones as well. You can read more about this topic on the Web: The Largest Known Primes: utm.edu/research/primes/largest.html.

2 Factoring Monomials

If the monomial you are factoring has no variables, then what you know about factoring numbers in arithmetic will be all you need. If you have variables, however, things can get a little trickier. You'll see how to deal with that sort of thing in this section.

Finding the GCF and LCM of Terms with Variables

Dear Dr. Math,

I really appreciate all of the help you have given me in the past. Here are some other questions I have; maybe you could help me out.

1. $(x^6)(xy^3)$. I understand the concept of exponents, but I don't understand what I must do in order to simplify this problem.
2. $(2rs^5)(-6mr^6)$. Same as above.
3. Find the greatest common factor: $12a^3c$ and $15ab^3$. How do you find the GCF of a number and two variables that have exponents?
4. Find the least common multiple: $7x$ and $8x^2$.

Last year in pre-algebra we never reviewed least common multiples, and I have forgotten how to find them.

Sincerely,

Aimee

Dear Aimee,

For your first problem: $(x^6)(xy^3)$

Use the associative and commutative properties to get the *x*'s and the *y*'s together by dropping parentheses and moving powers if necessary:

$$(x^6)(xy^3) = (x^6 \cdot x)(y^3)$$

Now use the distributive property for exponents, $n^a \cdot n^b = n^{(a+b)}$, remembering that $x = x^1$:

$$x^6 \cdot x = x^6 \cdot x^1 = x^{(6+1)} = x^7$$

Now put it all together:

$$x^7y^3$$

If your problem is in the use of the distributive property itself, it may help just to picture it in terms of repeated multiplication:

$$x^6 \cdot x \cdot y^3 = x \cdot x \cdot x \cdot x \cdot x \cdot x \quad \cdot x \quad \cdot y \cdot y \cdot y$$
$$= x \cdot x \cdot x \cdot x \cdot x \cdot x \cdot x \quad \cdot y \cdot y \cdot y$$
$$= x^7 \cdot y^3$$

It's really nothing more than counting factors.

Now let's look at your second problem:

$$(2rs^5)(-6mr^6)$$

This is similar to the first problem, except that you have to reorder the parts and there are numeric coefficients involved. First pull the problem apart by dropping the parentheses:

$$2\ r\ s^5\ (-6)\ m\ r^6$$

Then reorder (I like to put variables in alphabetical order to make sure I got them all):

$$2\ (-6)\ m\ r\ r^6\ s^5$$

Now you can combine numbers with numbers and powers of the same variable with one another as in the previous example:

$$-12\ m\ r^7\ s^5$$

and you're done.

Let's look at finding the greatest common factor of $12a^3c$ and $15ab^3$.

As with numbers, one good way to find the GCF is to list the factors of the two expressions in the same order (I like numerical order for primes and alphabetical order for variables), and then choose the largest power of each variable that will divide both—that is, the smallest power present in both rows of each column of my diagram:

Rx GCF AND LCM

GCF stands for Greatest Common Factor. For any two integers *a* and *b*, the GCF is the largest number that divides both *a* and *b* evenly.

LCM stands for Least Common Multiple. For any two integers *a* and *b*, the LCM is the smallest number that contains both *a* and *b* as factors.

$$12a^3c = 2^2 \cdot 3 \cdot a^3 \cdot c$$
$$15ab^3 = 3 \cdot 5 \cdot a^1 \cdot b^3$$
$$\text{GCF} = 3 \cdot a^1$$

The answer to the third problem is $3a$. The 3 is the GCF of 12 and 15, and the a is the only variable that is present in both expressions. The 2^2, 5, b^3, and c disappear because they each appear in only one of the expressions.

Your last problem is to find the least common multiple of $7x$ and $8x^2$.

Finding an LCM is the opposite of the GCF: list the factors, as above, but look for the *largest* power in either row of each column:

$$
\begin{array}{lcl}
7x & = 7 & \cdot \quad x \\
8x^2 & = & 2^3 \cdot x^2 \\
\hline
\text{LCM} & = & 7 \cdot 2^3 \cdot x^2 = 56x^2
\end{array}
$$

This is because the LCM has to contain at least the 7 and one x and also at least the 8 and two x's; that means it needs the 7, the 8, and two x's.

Have you noticed, by the way, that for any numbers or expressions a and b,

$$a \cdot b = \text{GCF}(a, b) \cdot \text{LCM}(a, b)$$

In this case, $a = 7x$, $b = 8x^2$. The GCF of these terms is x and the LCM is $56x^2$. The products on both sides are $56x^3$. That means you can easily find the LCM if you know the GCF. It works because the LCM includes all the factors of a and b, and the GCF consists of the factors that have to be included twice to form the complete product. If you think through this concept, it may help you understand better what's going on when you find the GCF or LCM. This picture illustrates it; see if you can follow what I mean:

$$
\begin{array}{rl}
a = & a_1\, a_2\, b_1\, b_2\, c_1\, c_2\, c_3 \\
b = & a_2\, b_1 \, c_2\, c_3\, d_1\, d_2 \\
\text{LCM} = & a_1\, a_2\, b_1\, b_2\, c_1\, c_2\, c_3\, d_1\, d_2 \\
\text{GCF} = & a_2\, b_1 \, c_2\, c_3
\end{array}
$$

—***Dr. Math, The Math Forum***

3 Factoring Polynomials

Remembering the skills that you use to factor a number will help as you move to the more complicated algebraic factoring. A prime number of objects can only be arranged into two possible rectangles (if you count, say, $1 \cdot 3$ and $3 \cdot 1$ as separate rectangles), and we can represent this with numbers or with pictures. For example, if we have the number 3:

A composite number, on the other hand, can be arranged into more than two possible rectangles. Let's look at the number 16:

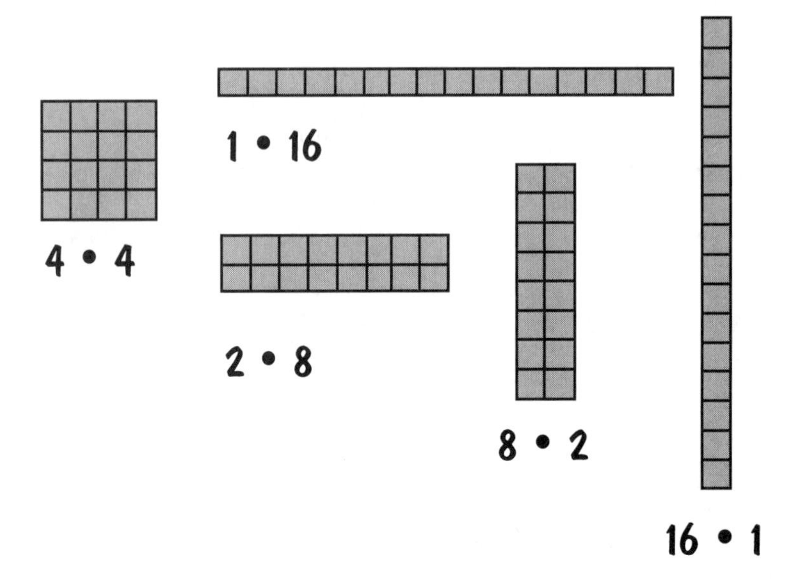

We can extend this idea to algebra. We can draw a rectangle that goes with this polynomial: $x^2 + 4x + 3$.

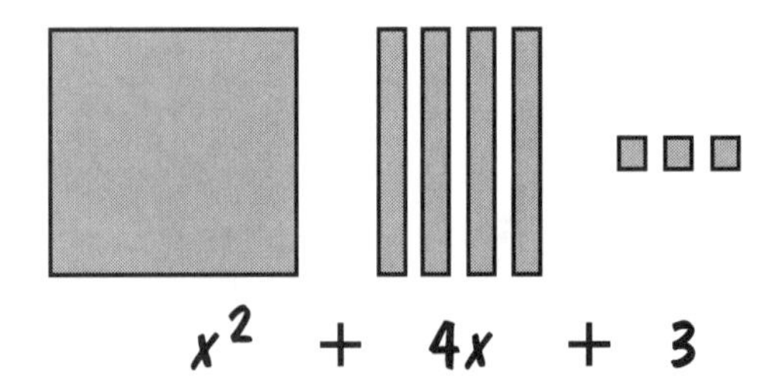

If you have algebra manipulatives to move around (either from a plastic set or made from paper), it's a little easier than just drawing. We identify each of the pieces by its area. If the large square has a side length of x, then the area of the large square is x^2. The skinny pieces are 1 unit in width and x units in length, so the area of each one is 1x. The small squares are unit squares. Each side length is a unit, and so the area is 1.

The puzzle is to move the different pieces so that you have a rectangle. If they fit, then you can look at the rectangle to see the factored form of the polynomial.

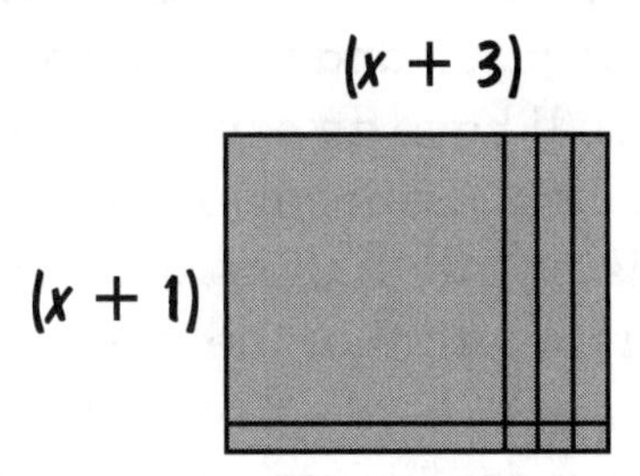

If the length is x + 1 and the width is x + 3, this picture represents the area of the rectangle. If we separate the parts and group them according to their shape and size, we have:

$$x^2 + 4x + 3 = (x + 3)(x + 1)$$

Understanding this representation can make factoring make more sense. And this section will give you some techniques or algorithms that will also help.

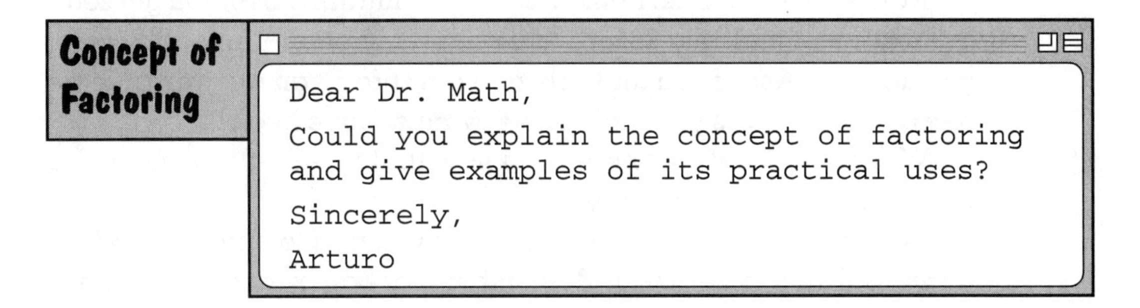

Concept of Factoring

Dear Dr. Math,

Could you explain the concept of factoring and give examples of its practical uses?

Sincerely,

Arturo

Dear Arturo,

Thanks for your question. Factoring is an idea you might be familiar with from multiplication. The numbers multiplied together to get another number are its factors.

For example, $4 \cdot 3 = 12$, so 3 and 4 are factors of 12. However, they're not its only factors. 1, 2, 6, and 12 are other factors of 12. (Another way of defining a factor is a number that divides evenly into the number you're factoring.)

The sort of factoring you're doing in algebra is somewhat similar, but there are probably letters like x and y stuck into the equations. These letters just stand for unknowns.

Sometimes you'll have an equation that has a squared or cubed term that is unknown. Here's an example: $x^2 = 9$.

Since the x is squared, there are *two* possible answers for x. You can probably guess that the answers will be -3 and 3, since $-3 \cdot -3 = 9$ and $3 \cdot 3 = 9$.

But suppose you don't know what the answer is (or suppose that you're dealing with a more complicated equation). How would you figure out what the answer is?

Here's where we get back to factoring. Here's what you'd do. First, subtract 9 from each side of the equation to get 0 on one side.

$$\begin{array}{r} x^2 = 9 \\ \underline{-9 \;\; -9} \\ x^2 - 9 = 0 \end{array}$$

Now, here's the trick. Remember, if you multiply $5 \cdot 0$, you get zero. In fact, if you multiply 28, or 1 billion, or any other number by zero, you get zero. And if you multiply and get zero as an answer, at least one of the numbers you multiplied by must have been 0!

We can use that fact here. We know that $x^2 - 9 = 0$. So one of the factors of $x^2 - 9$ must equal 0.

Next, we factor. Basically, this means saying to ourselves: What do we multiply to get $x^2 - 9$? Remember that both x^2 and 9 are *square* numbers.

From $x^2 - 9 = 0$, we get $(x + 3)(x - 3) = 0$.

Now we know that either $x - 3$ could be 0 or $x + 3$ could be 0.

If $x - 3 = 0$, then

$$\begin{array}{r} x - 3 = 0 \\ +3 \ \ +3 \\ \hline x = 3 \end{array}$$

If $x + 3 = 0$, then

$$\begin{array}{r} x + 3 = 0 \\ -3 \ \ -3 \\ \hline x = -3 \end{array}$$

—***Dr. Math, The Math Forum***

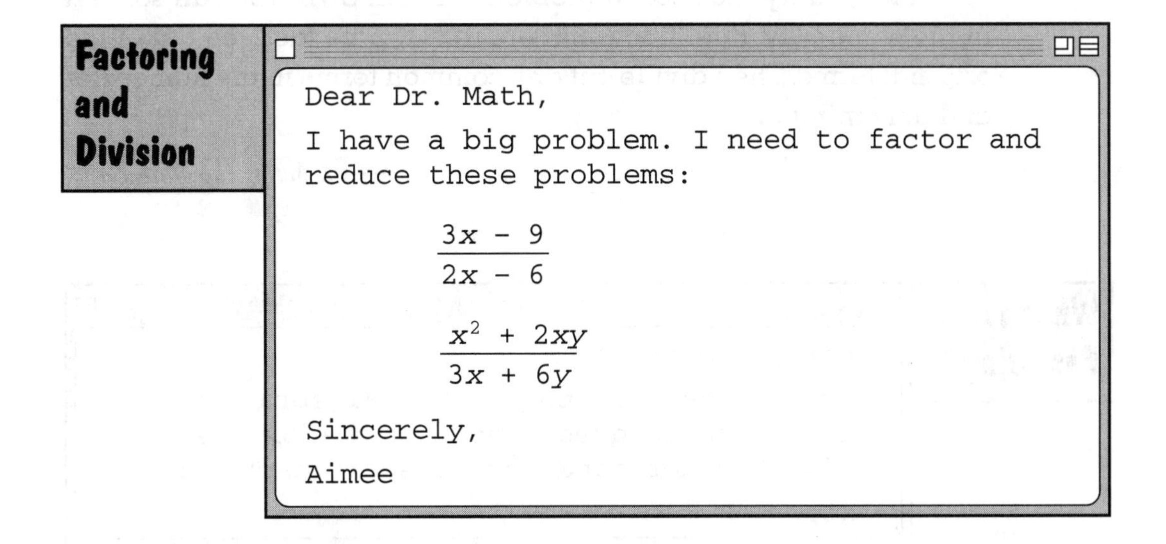

Dear Aimee,

When you want to solve problems like these, you need to look for two things. The first thing to do is to look for numbers that will divide all the terms within one expression and then a number that will divide the terms in the other expression. The second thing we need to do is to look at the possible factors (if there are more than one) and find ones that are common to both sides of the division sign.

For example, in the first problem you asked about, $(3x - 9)/(2x - 6)$, we can factor a 3 out of the first set of terms, since both 3x and 9 are divisible by 3. This gives us 3(x – 3). The second set of terms can be factored by 2, giving us 2(x – 3). Do you see how that works?

The problem now is

$$\frac{3(x-3)}{2(x-3)}$$

Do you see any common term we can cancel out?

How about (x – 3)? If we divide the (x – 3) in the numerator by the (x – 3) in the denominator, we get $\frac{3}{2}$, a much nicer number than what we started with.

I'll let you try the second problem on your own. You can solve it in the same way. First find numbers that can be factored out of the original terms. Then divide out any common terms in the numerator and denominator.

—Dr. Math, The Math Forum

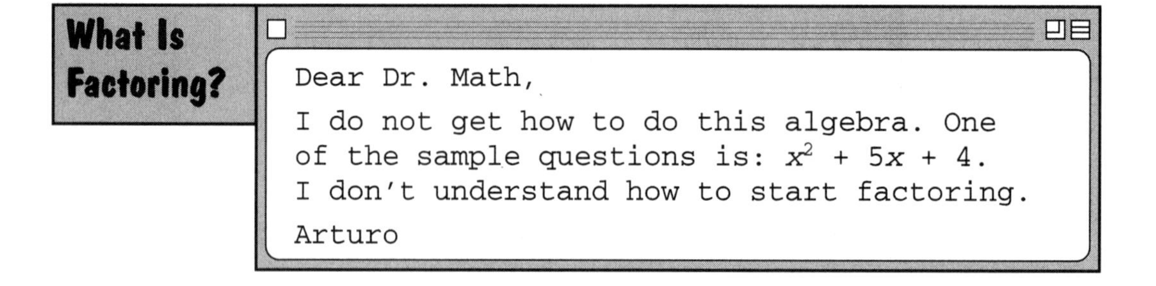

What Is Factoring?

Dear Dr. Math,

I do not get how to do this algebra. One of the sample questions is: $x^2 + 5x + 4$. I don't understand how to start factoring.

Arturo

Dear Arturo,

Factoring can be a very intimidating thing, but once you get the hang of it, you'll have a really useful skill.

First, let me make sure you understand what factoring is. To factor anything, you get it into its smaller component parts, finding out what smaller parts make up the bigger part. For instance, when you factor a number, say 12, you break it into its smallest parts. You could start with $6 \cdot 2$. Then 6 can be factored into $2 \cdot 3$, so your factors of 12 are $2 \cdot 2 \cdot 3$. So to factor $x^2 + 5x + 4$ is to break it down into simpler parts.

You may have already done some problems like this:

$(x + 3)(x - 4)$

Multiplying it out gives you:

$x^2 - 4x + 3x - 12$

Then simplify to get:

$x^2 - x - 12$

There are some very important things to notice here: You get the first term of the trinomial, x^2 (they call it a trinomial because three terms are being added) by multiplying the *first* terms in each parenthesis $(x \cdot x)$.

To get the middle term, first you multiply the first term in the first parenthesis (x) by the last term in the second parenthesis (–4) (we call these terms the *outside* terms because they are the outside terms of the two parentheses). Then you multiply the last term in the first parenthesis (3) by the first term in the second parenthesis (x) (we call these terms the *inside* terms because they are the inside terms of the two parentheses). Then add the "inside" and the "outside" products together. Thus the middle term becomes $3x - 4x = -x$.

The last number in the trinomial, the constant (meaning it has no x's in it), –12, comes from multiplying the *last* terms in each parenthesis $(3 \cdot -4)$.

You can remember this process by remembering the word FOIL, which stands for First, Outside, Inside, Last.

Factoring a trinomial is just doing the reverse of what I just did in the problem above. You want to get from the trinomial back to the factors. First make sure your problem is written in the right way: that is, the term with x^2 comes first (the first term), the term with x (also called the middle term) comes next, and the constant—the term with no x in it—comes last. In your problem, x^2 is the first term, 5x is the middle term, and 4 is the last term.

Step one is to write the problem on paper:

$x^2 + 5x + 4$

Step two is to write two empty parentheses like this:

$(\quad)(\quad)$

We do this because we know the factors of a trinomial look this way.

Step three is to look at the first term, which is x^2. We know from the problem I did above that the x^2 term comes from multiplying the first term in the first parenthesis by the first term in the second parenthesis. What are the only two things that can be multiplied together to give you x^2? That's right: $x \cdot x$ will give you x^2. You write it like this:

$(x \quad)(x \quad)$

Step four is to look at signs. Remember from the problem above that the constant term (the one with no x in it) is found by multiplying the last terms of each parenthesis. Because the constant term in your problem (4) is positive, the signs of both the numbers that are multiplied together to give you 4 have to be either positive or negative (because $+ \cdot +$ or $- \cdot -$ gives you a positive number, and $+ \cdot -$ will give you a negative number). So your parentheses could now look like this:

$(x - \quad)(x - \quad)$

or this:

$(x + \quad)(x + \quad)$

Look back at my previous problem to see what gave you the middle term. Remember, you are adding the "outside" to the "inside." If both signs are negative, this will give you a negative number; if both signs are positive, it will give you a positive number. Because $+5x$ is positive, you know that the signs have to be positive. Thus you get:

$(x + \quad)(x + \quad)$

Step five is the tricky part. You know that the last term is found by multiplying the last two numbers in each parenthesis together. This means the numbers could either be 4 and 1, or 2 and 2 (because $4 \cdot 1 = 4$, or $2 \cdot 2 = 4$). But you also know that the inside terms multiplied together plus the outside terms multiplied together will give you the middle term, so you have to try them out. Let's try 2 and 2:

$$(x + 2)(x + 2) = x^2 + 2x + 2x + 4 = x^2 + 4x + 4$$

Whoops! The middle term is not right. We need 5x, not 4x. Now let's try 4 and 1:

$$(x + 4)(x + 1) = x^2 + 4x + 1x + 4 = x^2 + 5x + 4$$

We got it. The factors of $x^2 + 5x + 4$ are $(x + 4)$ and $(x + 1)$.

Let's try another one: Factor $x^2 + x - 2$.

Here are our parentheses:

$$(\quad)(\quad)$$

Remember FOIL (First, Outside, Inside, Last): x^2 is the first terms of the parentheses multiplied together, and the only thing they can be is x and x:

$$(x \quad)(x \quad)$$

The sign in front of the constant is negative (–2), so the signs of the factors have to be different. This is the only way we can get a –2, found by multiplying the last terms.

$$(x + \quad)(x - \quad)$$

Now, because the term in front of the x term (+x) is a plus, we know that when we multiply the outside and inside terms together and add them to each other, we need to end up with a plus. Because the signs are different, the bigger term will have to be a plus. Now the only factors of 2 are 2 and 1, so they will be the only numbers we will have to worry about.

If we put 2 and 1 in like this:

$$(x + 2)\ (x - 1)$$

we can check this by multiplying it out:

$$x^2 - x + 2x - 2 = x^2 + x - 2$$

If we had reversed the 2 and the 1, we would have ended up with $-x$ instead of $+x$.

Sometimes it may take a little trial and error to get the right facts. Always multiply your answer back out to make sure you end up with something that works.

—Dr. Math, The Math Forum

Factoring Expressions

Dear Dr. Math,

I'm having problems with some of these questions. Can you please enlighten me on these?

I am not sure where I went wrong on this one. Can you please tell me?

$$4x^2 - 36 = a^2 - b^2$$
$$= (a + b)(a - b)$$
$$= (2x + 6)(2x - 6)$$

I've learned how to use the distributive law to show the following special rule:

$$(a + b)(a - b) = a^2 - b^2$$

But this one puzzles me:

$$x^2 + yz + xy + xz = ?$$

There are altogether four numbers. How does this work?

Sincerely,

Aimee

Dear Aimee,

I can see nothing wrong with what you have done with the first problem. I do it a little differently and put them into their "squared" form as soon as I can, that is:

$$4x^2 - 36 = (2x)^2 - (6)^2$$
$$= (2x - 6)(2x + 6)$$

The pattern that you cite, $(a + b)(a - b) = a^2 - b^2$, will help you a lot in algebra class.

As for the other one:

$$x^2 + yz + xy + xz$$

this is a matter of factoring four terms. But they're four terms with three variables repeated in them. When I get problems like this one, I always look to see if any variable is common to all four terms. Nope, not here: x is only in three of them. Well, can you group the terms by twos, with a common factor (variable) in each pair? Sure you can, like this:

$$x^2 + xy + yz + xz$$

Now you can factor the x out of the first pair of terms and the z out of the second pair of terms, like this:

$$x(x + y) + z(y + x)$$

Multiply it out if you are unsure, and you'll see that you end up with the original equation.

Then, notice that $(x + y)$ is the same as $(y + x)$. So, because both x and z are being multiplied by exactly the same thing, you can rewrite it as:

$$(x + z)(x + y)$$

To check that you have arrived at the correct answer, multiply your answer out; you should end up with the original expression. That way you know that you've found the right answer.

For students who worry that their factoring skills aren't up to par, it's good to remember what the mathematician Yitz Herstein said: "Factoring a quadratic becomes confused with genuine mathematical talent."

—Dr. Math, The Math Forum

Factoring the Difference of Two Squares

Sometimes there is a special case that can save you a lot of time! Here's what the difference of squares looks like using variables:

$$x^2 - y^2 = (x + y)(x - y)$$

The trick is to recognize a polynomial when it's in this form. Here are some that are:

$9x^4 - y^2$

$25x^2 - 49$

$x^6 - y^4$

One thing to notice is that all of them have no middle term and all involve subtraction. What about these, though?

$9x^4 + y^2$

$-49 + 25x^2$

If you remember what you're looking for, you can change these, without altering their value, into something that you can easily factor. Do you see how?

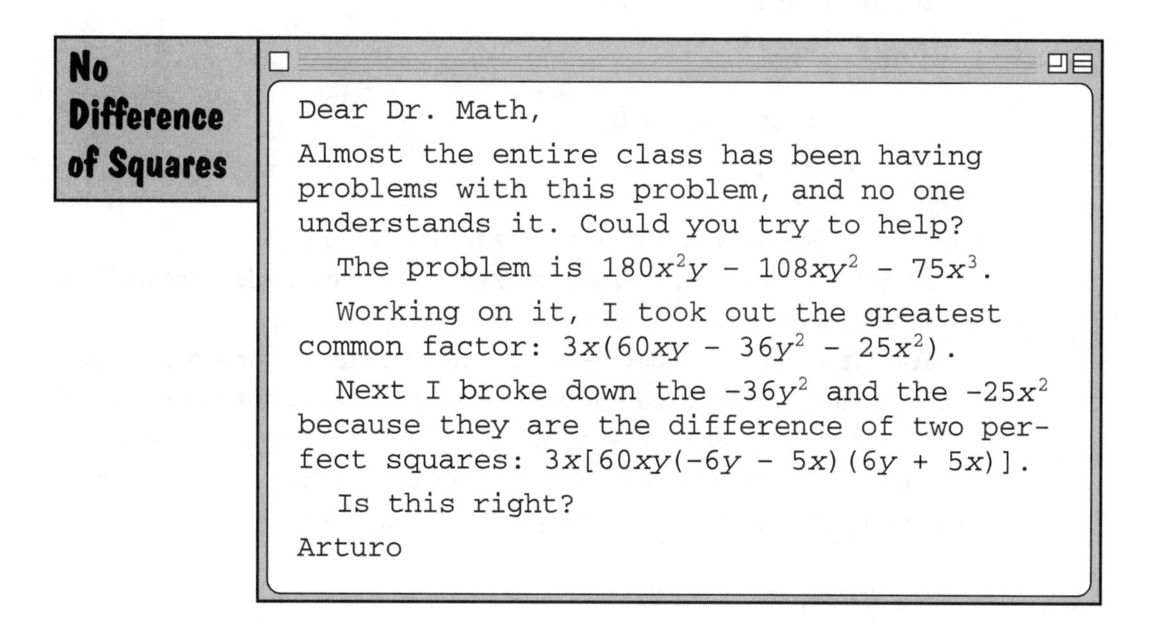

No Difference of Squares

Dear Dr. Math,

Almost the entire class has been having problems with this problem, and no one understands it. Could you try to help?

The problem is $180x^2y - 108xy^2 - 75x^3$.

Working on it, I took out the greatest common factor: $3x(60xy - 36y^2 - 25x^2)$.

Next I broke down the $-36y^2$ and the $-25x^2$ because they are the difference of two perfect squares: $3x[60xy(-6y - 5x)(6y + 5x)]$.

Is this right?

Arturo

Dear Arturo,

Your first step is correct:

$$3x(60xy - 36y^2 - 25x^2)$$

Next you looked at $-36y^2 - 25x^2$ and tried to factor it as a difference of squares. But this is not a difference of squares, because $-36y^2$

is not a square (unless you're getting into imaginary numbers, which probably isn't the case here). $36y^2$ is a square, however. Just check: $(-6y - 5x)(6y + 5x) = -36y^2 - 60xy - 25x^2$, which is not the same as $-36y^2 - 25x^2$.

Factoring $-36y^2 - 25x^2$ doesn't help you much. Suppose, for argument's sake, that factoring $-36y^2 - 25x^2 = (-6y - 5x)(6y + 5x)$ were correct. You still wouldn't have $3x[60xy(-6y - 5x)(6y + 5x)]$. Instead, you would have $3x[60xy + (-6y -5x)(6y + 5x)]$. This doesn't help much, since you need a factoring that includes the $60xy$.

But you are on the right track! It is important that $36y^2$ and $25x^2$ are both squares. Since there is no difference of squares (and you need to involve the $60xy$ anyway), you can't use the rule for difference of squares. What you need to learn to factor is something that is *itself* a square.

Look at these two products:

$$(a + b)^2 = a^2 + 2ab + b^2$$
$$(a - b)^2 = a^2 - 2ab + b^2$$

(You can check these for yourself by multiplying.)

Notice that there are squares in these formulas also, as well as $2ab$, the *cross term* in this case.

You have $3x(60xy - 36y^2 - 25x^2)$. You can't use your squares yet, because they're negative squares right now. So factor out -1 to get $-3x(-60xy + 36y^2 + 25x^2)$, or $-3x(36y^2 - 60xy + 25x^2)$. Do you see that the expression in parentheses here matches one of the patterns above? Following that, you'd finish the problem this way:

$$-3x(36y^2 - 60xy + 25x^2)$$
$$-3x(6y - 5x)(6y - 5x)$$
$$-3x(6y - 5x)^2$$

If you check by multiplying this out to expanded form, you should get the original expression.

—Dr. Math, The Math Forum

Difference of Squares of Integers

Dear Dr. Math,

I had this question at school: Which positive integers can be written as the difference of the squares of two integers? Explain your reasoning.

I know that the numbers that are squared cannot be equal, so I came up with this: $n^2 - (n - x)^2$ = a positive integer.

The smallest number that will work is 3: $2^2 - 1^2 = 3$. Continuing sequentially, the differences turn out to be odd numbers. If I try nonsequential numbers, however, the difference can be even: $6^2 - 4^2 = 20$.

I have an equation, but I don't know how to make it work. Can you help?

Sincerely,

Aimee

Dear Aimee,

Note that for $n^2 - (n - x)^2$ to equal a positive integer, n and x must be positive integers.

Any expression of the form $a^2 - b^2$ can be factored as $(a - b)(a + b)$. It follows that if a number can be factored in this way, then it can be expressed as a difference of squares.

The example you gave could be worked as follows:

$$20 = 10 \cdot 2$$

So: $a + b = 10$

$\underline{a - b = 2}$

Add: $2a = 12$ so $a = 6$

Subtract: $2b = 8$ so $b = 4$

You will note that the sum of the factors must be even, since they will equal $2a$ (and the difference will equal $2b$). So if a number can be factored into two even or two odd factors, then it can be expressed as a difference of squares.

Example: $35 = 7 \cdot 5$ $\quad a + b = 7$

$a - b = 5$

$2a = 12$ and $a = 6, b = 1$

And it is true: $35 = 36 - 1 = 6^2 - 1^2$.

Example: $63 = 9 \cdot 7$ $\quad a + b = 9$

$a - b = 7$

$2a = 16, a = 8, b = 2$, and $8^2 - 1^2 = 63$

Example: $60 = 10 \cdot 6$ $\quad a + b = 10$

$a - b = 6$

$2a = 16, a = 8, b = 2$, and $8^2 - 2^2 = 60$

So, the general rule is that, provided a number can be factored with two even or two odd factors, it can be represented as a difference of squares.

POLYNOMINAL FACTORING:

To give you a quick summary of polynomial factoring:

1. **Always remove common factors first.**
2. **If it has two terms look for:**
   - difference of two squares
   - sum or difference of two cubes (if a cubic)
3. **If it has three terms:**
   - it may be a perfect square (then use $(a + b)^2$ or $(a - b)^2$)
   - it may be factored by using the distributive law in reverse
   - try completing the square
4. **If it has four or more terms it may be factored by:**
   - grouping the terms in order to apply other methods

—Dr. Math, The Math Forum

Here's how to show graphically that $(a^2 - b^2) = (a + b)(a - b)$:

a

a

b

b

=

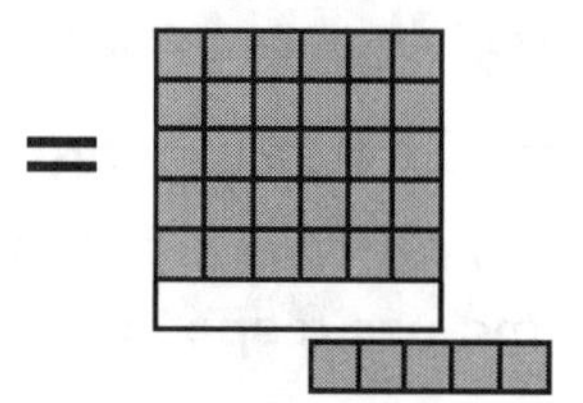

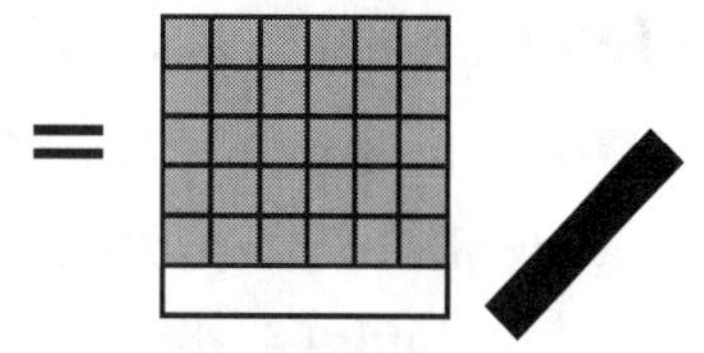

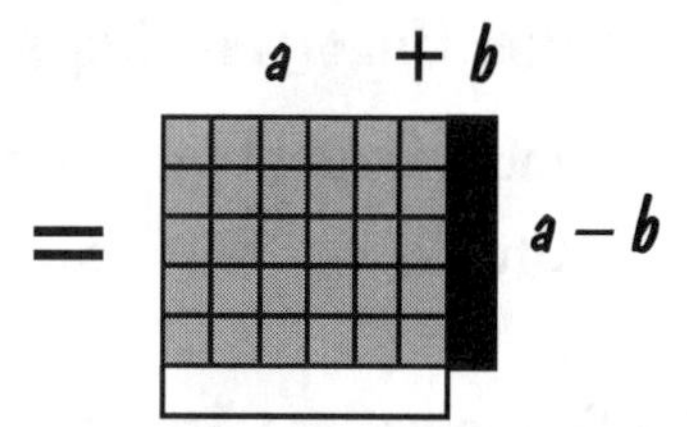

Learn more about factoring at these sites:

Math Forum: Algebra Problem of the Week: All Wet

mathforum.org/algpow/solutions/solution.ehtml?puzzle=35

Solve simultaneous equations, find area, and use the Pythagorean theorem and percentages to find the spraying distance needed for a circular sprinkler in a rectangular yard.

Math Forum: Algebra Problem of the Week: A Product-Plus-1

mathforum.org/algpow/solutions/solution.ehtml?puzzle=73

The product of any four consecutive integers, increased by one, is always a square number. Give at least three instances of that statement and prove that this will always occur by finding an algebraic expression for that "square number."

Math Forum: Algebra Problem of the Week: Trick-or-Treat Box

mathforum.org/algpow/solutions/solution.ehtml?puzzle=89

Harvey and his classmates make trick-or-treat boxes in math class and use a special formula to find the volumes.

Math Forum: Algebra and Calculus Sketches

mathforum.org/sum95/ruth/sketches/algcalc.sketches.html

Exploring equations for lines, parabolas, and tangents using the Geometer's Sketchpad.

Math Forum: Understanding Algebraic Factoring

mathforum.org/alejandre/algfac.html

Students use algebra tiles to explore algebraic factoring.

Shodor Organization: Project Interactivate: Coloring Multiples in Pascal's Triangle

shodor.org/interactivate/activities/pascal1/

Students color numbers in Pascal's Triangle by rolling a number and then clicking on all entries that are multiples of the number rolled, thereby practicing multiplication tables, investigating number patterns, and investigating fractal patterns.

Quadratic Equations

A **quadratic equation** is an assertion about a single variable. For example,

$$-1 = 3x^2 - 5x$$

A **quadratic function** is what we get when we introduce a second variable, which allows for a whole range of solutions:

$$y = 3x^2 - 5x + 1$$

If we graph the individual solutions of a quadratic function, we get a parabola. In many cases, we want to know where the parabola crosses the x-axis (called the "roots" of the function). We can find the roots by setting the value of *y* to zero, and solving the resulting quadratic equation for x:

$$0 = 3x^2 - 5x + 1$$

Sometimes we can use factoring to solve this problem, but in general, that just won't work. However, mathematicians have been nice enough to work out the **quadratic formula,**

$$x = \frac{-b \pm \sqrt{b^2 - 4ac}}{2a} \quad \text{whenever } ax^2 + bx + c = 0$$

which we can use to find the roots of *any* quadratic function.

In this part, Dr. Math explains:

- quadratic expressions, equations, and functions
- solving quadratic equations by factoring
- solving quadratic equations by graphing
- solving quadratic equations by taking square roots
- solving quadratic equations by completing the square
- the quadratic formula

1 Quadratic Expressions, Equations, and Functions

Quadratic functions, equations, and expressions all have one thing in common: The term with the highest power has an exponent equal to 2.

A quadratic *function* will look like

$$f(x) = ax^2 + bx + c$$

or

$$y = ax^2 + bx + c$$

depending on which notation is being used. Sometimes you'll see them combined:

$$y = f(x) = ax^2 + bx + c$$

If we set the value of the function to be a constant, we get a quadratic *equation,* for example,

$$3 = ax^2 + bx + c$$

Usually the constant will be zero; but even if it's not, we can always arrange to have zero on one side by itself:

$$0 = ax^2 + bx + \underbrace{(c - 3)}_{\uparrow \text{ our new "c"}}$$

Without an equal sign, we're left with a quadratic *expression:*

$$ax^2 + bx + c$$

whose value depends on whatever value we decide to assign to the variable.

QUADRATIC FUNCTIONS

An equation is any "number sentence" that says two expressions are equal.

A function is a relation between two or more variables, such that for any value of the **independent variable(s)**, there is exactly one value for the function. Sometimes it helps to think of a function as a machine where you input data, and because the data go through the function, you receive output. This idea of a machine helps us think about what the difference is between an independent variable (the input) and a **dependent variable** (the output—"dependent" because it depends on the input).

The function itself can commonly be expressed by a name, such as "f"; and it may be defined by stating the function's value as equal to an expression:

$$f(x) = 2x + 2$$

This is read: "f of x equals two x plus two." Note that the function is f, not $f(x)$, and the $f(x)$ is simply the value of the function for a given value of x. The equation is also not the function; the equation is being used to tell us the value of the function for any x, and thereby to define the function itself.

Not all functions can be expressed this way; for example, the square root function can't be expressed other than by the square root symbol. We would define that function by saying something like:

$\sqrt{x}$ is the nonnegative number y for which $y^2 = x$.

Here the equation $y^2 = x$ doesn't fully define the function by itself. Say you are given an equation relating variables x and y and asked whether this defines a function of x. You would solve the equation for y and see if this can be done in such a way that for every x, only one value of y will satisfy the original equation. In the case of $y^2 = x$, this can't be done; but by restricting y to nonnegative values, we can define a function.

●●

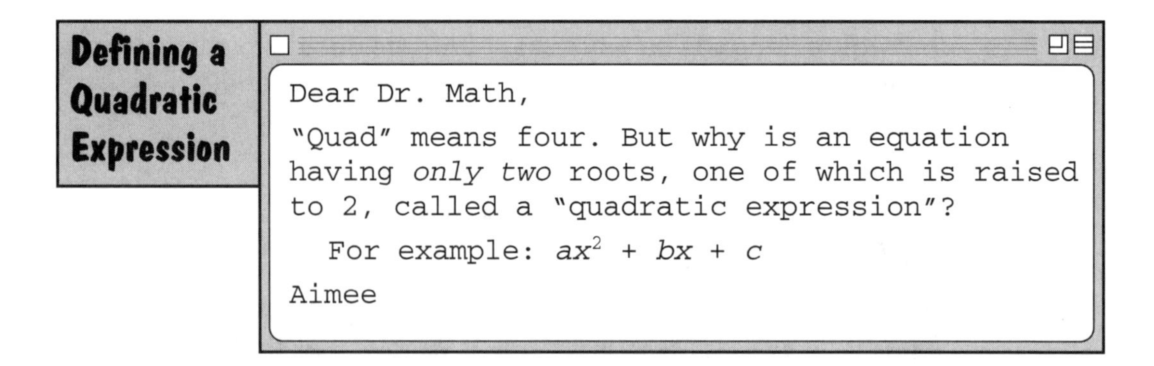

Defining a Quadratic Expression

Dear Dr. Math,

"Quad" means four. But why is an equation having *only two* roots, one of which is raised to 2, called a "quadratic expression"?

For example: $ax^2 + bx + c$

Aimee

Dear Aimee,

People often wonder about the word "quadratic," because they know that "quad" usually means four, yet quadratic equations involve the second power, not the fourth. But there's another dimension to the word.

Although in Latin the prefix "*quadri-*" means four, the word "*quadrus*" means a square (because it has four sides) and "*quad-*

ratus" means squared. We get several other words from this: "quadrille," meaning a square dance; "quadrature," meaning constructing a square of a certain area; and even "square" (through French).

Quadratic equations originally came up in connection with geometric problems involving squares, and of course the second power is also called a "square," which accounts for the name. The third-degree equation is similarly called a "cubic," based on the shape of a third power. Then when higher-degree equations began to be studied, the names for them were formed differently, based on degree rather than shape (since the Romans had no words for higher-dimensional shapes), giving us the quartic, quintic, and so

on. In fact, "quartic" came along later; originally, a fourth-degree equation was called "biquadratic," meaning doubly squared, which mixes the two concepts and is doubly confusing.

Here's a table of names for polynomials and their sources:

Degree	Name	Shape	Dimension
1	linear	line	1
2	quadratic	square	2
3	cubic	cube	3
4	quartic	—	4
5	quintic	—	5

—Dr. Math, The Math Forum

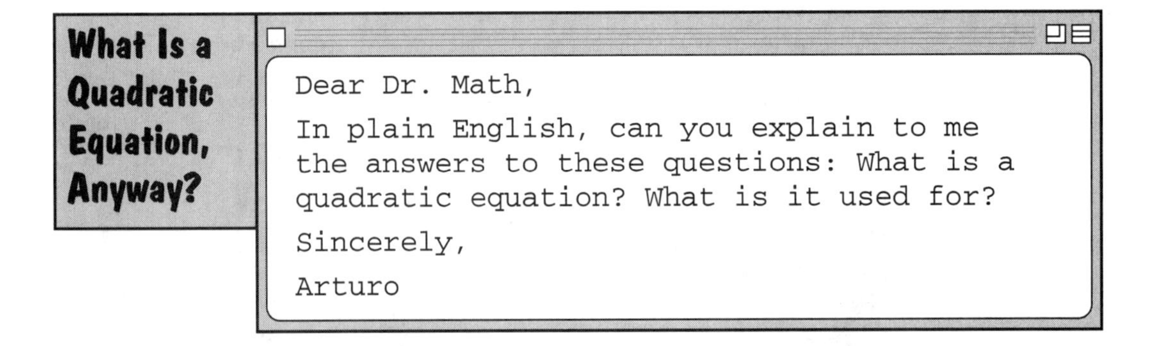

Dear Arturo,

Let's look at two examples:

> I'm thinking of two numbers. Their sum is 10 and their product is 21. What are the two numbers?

To solve this, you will end up naturally with a quadratic equation: Let the two numbers equal x and $10 - x$. Then their product is $x(10 - x) = 21$. Simplifying, you get:

$$x(10 - x) = 21$$
$$10x - x^2 = 21$$
$$-10x + x^2 = -21$$
$$x^2 - 10x + 21 = 0$$

or generally, for any two numbers that have sum s and product p, we can say that

$$x^2 - sx + p = 0$$

There generally will be two solutions to a quadratic equation corresponding to the two numbers.

In fact, all quadratic equations essentially come about in this manner. People consider the equation

$$ax^2 + bx + c = 0$$

to be the general form of the quadratic equation, where a, b, and c are constants with a not equal to zero. That doesn't look much like $x^2 - sx + p = 0$, does it? But if you divide this through by a, you will get an equation looking like $x^2 - (\frac{-b}{a})x + \frac{c}{a} = 0$. That's the same as our $x^2 - sx + p = 0$, if you take $s = \frac{-b}{a}$ and $p = \frac{c}{a}$.

Here's the second example:

> A square picture frame contains a picture with a mat border. The border is 3 inches thick on the sides and 4 inches thick on the top and bottom. If the area exposed within the mat border is 528 square inches, what are the dimensions of the original frame?

Again, a quadratic equation will arise naturally: The outside border of the mat is square, so let each side length equal x. Then the inner horizontal border is x minus 3 inches on one side and 3 inches on the other, or $x - 6$. Likewise, the inner vertical border is x minus 4 inches at the top and 4 inches at the bottom, or $x - 8$. If the inner area exposed measures 528 square inches, then

$$(x - 8)(x - 6) = 528$$
$$x^2 - 14x + 48 = 528$$
$$x^2 - 14x - 480 = 0$$

Other places where a quadratic equation may surface come from geometry, such as trying to find the intersection of a line and a circle, and in physics, such as when studying how objects fall to Earth.

—Dr. Math, The Math Forum

DEMONSTRATING A PATH

If you have access to a blackboard, you can toss a ball and note that the ball follows a path that looks like the graph of a quadratic function. Draw on the blackboard, as accurately as you can, a graph of the quadratic function $f(x) = \frac{-x^2}{36}$. Draw it so that you get a nice section of it on the board like this:

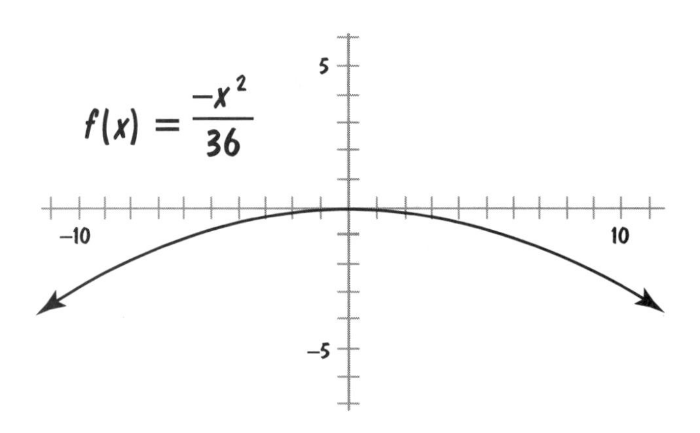

With practice, you can toss a ball across the blackboard so that it exactly follows the graph you have drawn. It's pretty exciting to see this done.

Quadratics in the Real World

Dear Dr. Math,

I'm wondering how quadratics help us in the real world.

Aimee

Dear Aimee,

There are lots of ways that quadratics are used in the real world. For example, NASA uses quadratics when they hurl rockets into space.

One example of a quadratic is a parabola. The term with the highest power in the equation of a parabola has an exponent equal to 2. An example of a quadratic that is a parabola is:

$y = c + b \cdot x + ax^2$ Notice the exponent.

If we use different letters we get:

$$y = D + V \cdot t + \frac{A}{2}t^2$$

which is the equation for the distance that something will move vertically when rising against gravity, where

y is the vertical distance

t is the length of time

D is the starting distance (perhaps 0)

V is the starting vertical velocity (perhaps 0)

A is the acceleration

So if you wanted to know how high a rocket would go (to determine if it would make it to outer space), you would use a quadratic like that one.

—***Dr. Math, The Math Forum***

2 Solving Quadratic Equations by Factoring

When factoring any quadratic equation, you are usually looking for two numbers that will:

1. multiply together to give the end number in the expression and
2. add/subtract together to give the middle number

This applies for any quadratic polynomial as long as the coefficient (the number in front) of x^2 is 1.

Factoring, once you're good at it, is just about the easiest way to solve quadratic equations.

Using Coefficients in Factoring

Dear Dr. Math,

How do I solve: $x^2 + x - 12 = 0$?

Arturo

Dear Arturo,

The way that most people like to solve problems like this is to factor the expression $x^2 + x - 12$. I'll do an example that's similar to yours, then you can use my method to do your problem. I'll use $x^2 + x - 30 = 0$.

First, figure out all the ways you can factor 30 as the product of two numbers:

$1 \cdot 30$

$2 \cdot 15$

$3 \cdot 10$

$5 \cdot 6$

Then, since we're going to end up with something in the form of (x – one factor)(x + the other factor)—remember why?—we use this to decide which of the pairs of factors we should use. At this point, it looks like guess-and-check, but there's another piece of information that will help us "guess" the right answer.

The key here is that we're looking for factors that have a difference of 1. That's because the coefficient of the x term is 1. So looking down the list, it looks like $5 \cdot 6$ will work. Should we use $(x - 5)(x + 6)$ or $(x - 6)(x + 5)$? Well, the first one gives us $x^2 + x - 30$, and the second gives us $x^2 - x - 30$. So we use the first one.

Once we have that, we know that x is either 5 or –6.

Do you see how you might do your problem now? Factoring is something that really takes a lot of practice to get good at.

—Dr. Math, The Math Forum

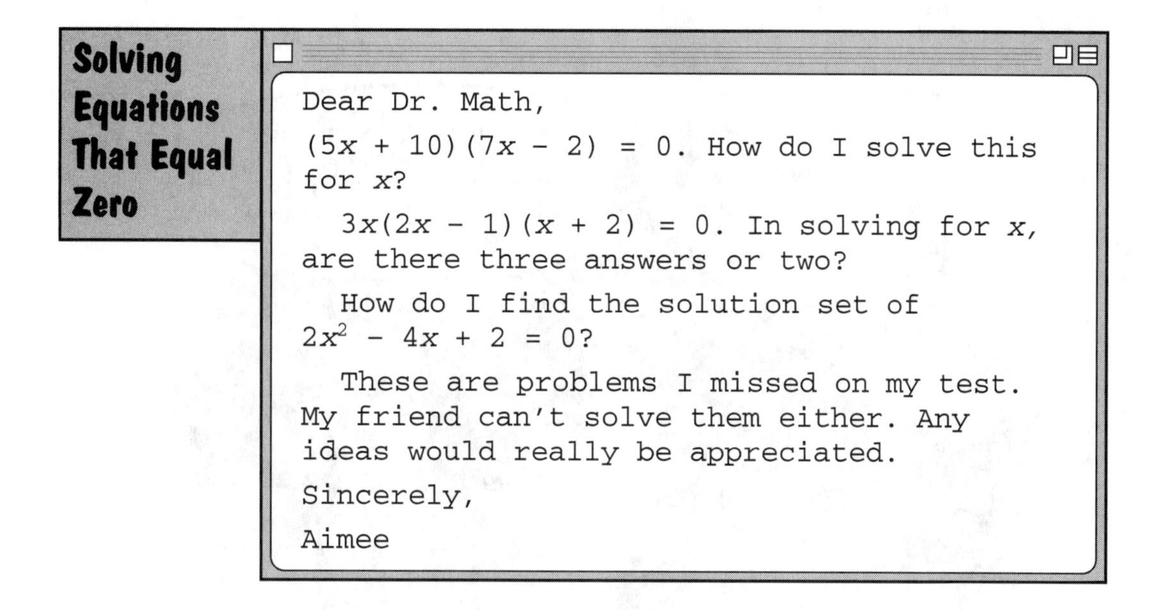

Solving Equations That Equal Zero

Dear Dr. Math,

$(5x + 10)(7x - 2) = 0$. How do I solve this for x?

$3x(2x - 1)(x + 2) = 0$. In solving for x, are there three answers or two?

How do I find the solution set of $2x^2 - 4x + 2 = 0$?

These are problems I missed on my test. My friend can't solve them either. Any ideas would really be appreciated.

Sincerely,

Aimee

Dear Aimee,

There is a key fact that should help you here. If a product of two numbers (two numbers multiplied together) is equal to zero, then at least one of them must be zero. In the case of your first problem, this means that either $5x + 10 = 0$ or $7x - 2 = 0$. The solutions to these two equations are $x = -2$ and $x = \frac{2}{7}$, respectively, so these are the solutions to the original equation. Your second problem can be solved in a similar fashion. There will be three answers, because there are three different factors.

For problems like your third one, where the equation is not already factored as in your first two problems, it is best to make the equation as simple as possible before we start to factor. In this case, since all the terms on the left side have an even number as their coefficient, we can factor 2 out of everything to get:

$$2(x^2 - 2x + 1) = 0$$

This factors as

$$2(x - 1)(x - 1) = 0$$

Since both of these factors are the same, there is only one answer, namely $x = 1$.

—Dr. Math, The Math Forum

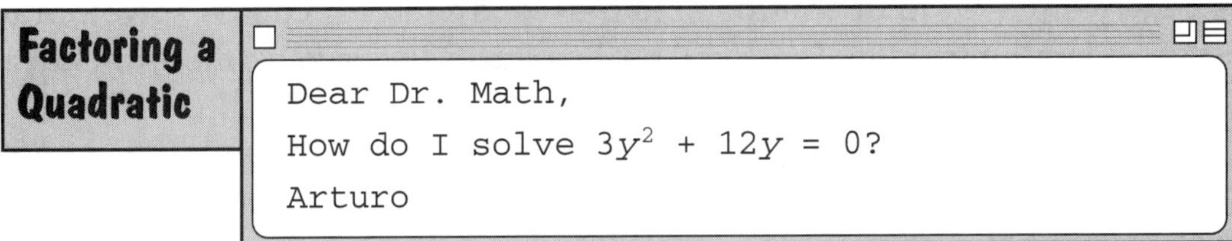

Factoring a Quadratic

Dear Dr. Math,

How do I solve $3y^2 + 12y = 0$?

Arturo

Dear Arturo,

You have probably seen the distributive property. One of several versions of it is:

$$(a + b) \cdot c = a \cdot c + b \cdot c$$

You can apply this to the left side of your equation, because y^2 is really y times y. It becomes:

$$(3y + 12) \cdot y = 0$$

We have not changed the equation much; just factored the left side.

Now what? In order for y to be a solution to this equation, one of the two things multiplied on the left must be equal to zero. That means, from the first factor, $y = 0$ will work. But how do you choose y so that $3y + 12 = 0$? This is a simpler kind of equation, which you have seen before. Solving that equation for y will give you the other solution.

Hint: the second solution is a negative number. Good luck.

—***Dr. Math, The Math Forum***

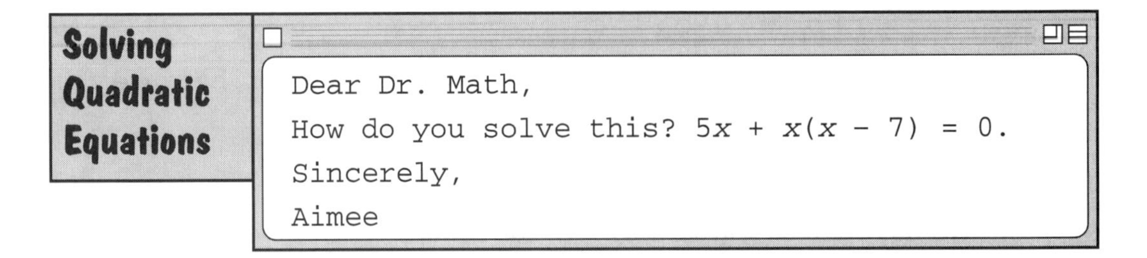

Solving Quadratic Equations

Dear Dr. Math,

How do you solve this? $5x + x(x - 7) = 0$.

Sincerely,

Aimee

Dear Aimee,

The first thing to do to solve any polynomial is to make sure it's in simplest form. For this one, we should start by multiplying everything out:

$$5x + x(x - 7) = 0$$

From the distributive property of multiplication, we can say that $x(x - 7) = (x \cdot x) - (x \cdot 7)$, which becomes $x^2 - 7x$.

$$5x + x^2 - 7x = 0$$

Now we can take the plain old "x" parts and add them:

$$x^2 + 5x - 7x = 0$$
$$x^2 - 2x = 0$$

Okay. Now that we've multiplied everything out and added it together, we have to factor again! Both expressions have an "x," so let's factor out the "x" so that we have:

$$x(x - 2) = 0$$

Now we have "something" times "something else" equals zero. In order for the equation to equal zero, either the "something" or the "something else" has to equal zero, so there are two possible answers for x:

if $x = 0$, we have $0(0 - 2) = 0$, which is true because $0 \cdot -2$ does equal 0.

if $(x - 2) = 0$, we have $x = 2$, so $2(2 -2) = 0$, which also works.

So $x = 0$ and $x = 2$ are our possible answers.

—Dr. Math, The Math Forum

Factoring Polynomials

Dear Dr. Math,

I'm very confused on how to go about factoring polynomials. I know, for instance, if you have a problem like this, you change it into two parentheses to figure it out. But then I get stuck; please help me.

Arturo

Dear Arturo,

It probably seems like magic unless you know why anyone cares about being able to do this. Suppose you have an equation like:

$$y = x^2 - 5x + 6$$

and you'd like to plot the equation on a graph. You could start putting in values of x and getting corresponding values of y:

$$x = 1, y = 1^2 - 5(1) + 6 = 2$$
$$x = 2, y = 2^2 - 5(2) + 6 = 0$$

and so on. But how do you know which values to choose for x?

It turns out that many quadratic equations can be written in a different form, which looks like

$$y = (x - 2)(x - 3)$$

What's the *point* of writing it this way? Well, note that right away we can find the places where the graph of the equation crosses the x-axis. How? By setting $y = 0$:

$$0 = (x - 2)(x - 3)$$

Now, the *only* way that you can multiply two things together and get zero is if at least one of them is zero. So we know that the only values of x for which this equation can be true are:

$$x = 2, 0 = (2 - 2)(2 - 3)$$

or

$$x = 3, 0 = (3 - 2)(3 - 3)$$

At *every* other value for x, you get something besides zero.

This tells us the locations of the points on the graph where the equation crosses the x-axis: (2, 0) and (3, 0).

We also know that the graph of a quadratic equation is a parabola, and since parabolas are symmetrical, that means that if the parabola crosses the line at these points, then the vertex of the parabola—the lowest or highest point, depending on whether it opens up or down—must be exactly halfway between these, at $x = 2.5$.

To find the location of that point, we substitute this value into the equation for *x:*

$$\begin{aligned} y &= (2.5 - 2)(2.5 - 3) \\ &= (0.5)(-0.5) \\ &= -0.25 \end{aligned}$$

So now we know where the vertex is: (2.5, –0.25).

From these three points, it's easy to sketch the graph. That's one reason we want to be able to factor equations this way. Here is another:

Sometimes a problem leads to an equation like:

$$y = \frac{x^2 - 5x + 6}{x^2 + 2x - 8}$$

Now this is pretty messy. But note that we can factor both the numerator and denominator:

$$y = \frac{(x - 2)(x - 3)}{(x - 2)(x + 4)}$$

$$\begin{aligned} y &= \frac{(x - 2)}{(x - 2)} \cdot \frac{(x - 3)}{(x + 4)} \\ &= \frac{(x - 3)}{(x + 4)} \end{aligned}$$

and get a much nicer equation to deal with.

The point I'm trying to make is that factoring quadratic (and other) equations isn't just something that math teachers made up to torture math students. If you can factor a quadratic, it becomes much easier to deal with. You can't always do it, but if you can, it's almost always a good idea.

Now the reason I explained all that wasn't just to test your ability to stay awake, but to try to give you enough context about what's going on so that if you *forget* how to factor an equation, you'll be able to figure it out again from scratch.

The basic idea is a very general one, which pops up over and over again in mathematics. To change something from one form to another, you write it in the other form using parameters; and then you figure out what the parameters have to be.

In this case, we know that we want to end up with something that looks like $(x + a)(x + b)$. So we go ahead and pretend that we've already found it:

$$x^2 - x - 12 = (x + a)(x + b)$$

Now we can expand the right side, to get something in the same form as the left side, but using our new parameters (a and b):

$$\begin{aligned} x^2 - x - 12 &= (x + a)(x + b) \\ &= x^2 + (a + b)x + ab \end{aligned}$$

Now this is pretty interesting, because if these are really equal, then they have to have the same coefficients:

$$\underline{\qquad ab = -12 \qquad}$$

$$\downarrow \qquad\qquad\qquad \downarrow$$

$$x^2 - x - 12 = x^2 + (a + b)x + ab$$

$$\uparrow \qquad\qquad\qquad \uparrow$$

$$\overline{\qquad\qquad\qquad\qquad}$$

$$a + b = -1$$

This gives us two constraints:

1. $a + b = -1$
2. $a \cdot b = -12$

This doesn't *look* like much of an improvement, but actually, we have a third constraint that turns out to be very helpful:

3. a and b are both integers

How does that help? Well, it turns out that there just aren't that many ways to multiply two integers to get –12. In fact, here are all the possibilities:

a	b
–1	12
1	–12
–2	6
2	–6
–3	4
3	–4

And if we check the sums, we'll find that only one pair of values will add up to –1:

a	b	$a + b$	
–1	12	11	
1	–12	–11	
–2	6	4	
2	–6	–4	
–3	4	1	
3	–4	–1	The winner!

Now we know that, if we did everything correctly, the factored version of the equation must be:

$$\begin{aligned} x^2 - x - 12 &= (x + 3)(x + -4) \\ &= (x + 3)(x - 4) \end{aligned}$$

We always want to check to make sure that we didn't make a mistake. In this case, we do that by plugging –3 and 4 into the original equation:

$$-3^2 - (-3) - 12 = 9 + 3 - 12$$
$$= 0$$
$$4^2 - (4) - 12 = 16 - 4 - 12$$
$$= 0$$

When it works, the key to finding the values of *a* and *b* is to narrow down the possibilities using the fact that $a \cdot b$ has to be equal to the constant at the end. When you get some practice at this, you'll be able to do it in your head without even writing anything down. For example, suppose you see:

$$x^2 - 6x + 8$$

The first thing you look at is the sign of the final term. It's positive, and the only way that you can get a positive term is by multiplying two positives or two negatives. So the possibilities are:

$1 \cdot 8$

$-1 \cdot -8$

$2 \cdot 4$

$-2 \cdot -4$

and the only pair that adds up to –6 is the final pair. So the answer must be:

$$(x - 2)(x - 4)$$

If the final constant is negative, for example,

$$x^2 + 2x - 15$$

then the factors have to have opposite signs:

$-1 \cdot 15$

$1 \cdot -15$

$-3 \cdot 5$

$3 \cdot -5$

As always, only one pair—the third one—gives the right sum, so the answer must be:

$$(x - 3)(x + 5)$$

So what at first seems to be a big, messy process boils down to a very simple shortcut at the end. However, if you just try to memorize the shortcut, you're likely to forget it under pressure—for example, during a test. But if you understand *why* the shortcut works, then you can work it out from scratch whenever you need it.

—Dr. Math, The Math Forum

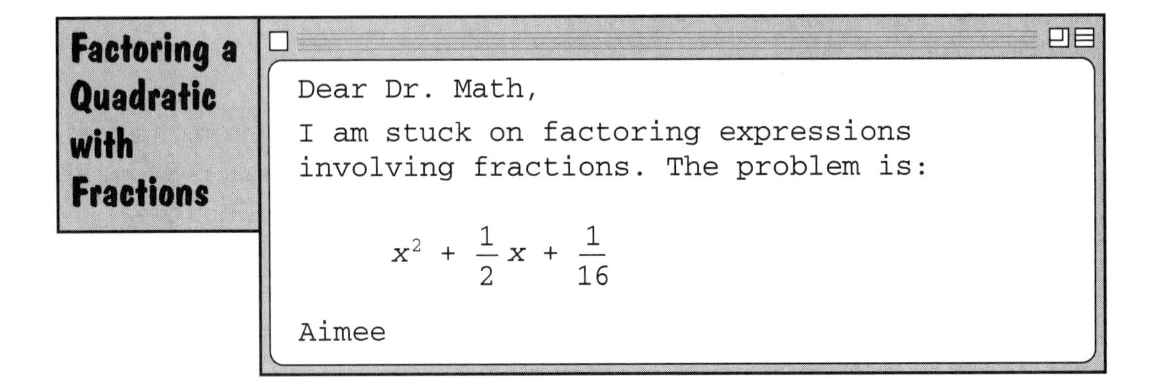

Factoring a Quadratic with Fractions

Dear Dr. Math,

I am stuck on factoring expressions involving fractions. The problem is:

$$x^2 + \frac{1}{2}x + \frac{1}{16}$$

Aimee

Dear Aimee,

Let's see if I can help you.

When factoring *any* quadratic, you are looking for two numbers that will:

1. multiply together to give the end number in the expression and
2. add/subtract together to give the middle number.

This applies for any quadratic polynomial as long as the coefficient (the number in front) of x^2 is 1, which in this case it is.

So, what numbers multiply to give $\frac{1}{16}$ and add (in this case) to give $\frac{1}{2}$? As you do this, you have to keep your eye on the signs that you want to end up with as well. In this case, you have only positive terms, so there are no problems with that.

For now, let's think of common factors of 16 and 2 (the two denominators of our fractions). What about the number 4? $4 \cdot 4 = 16$ and $2 \cdot 4 = 8$. If we take $\frac{1}{4}$ so that the 4 is also a denominator, then we have:

$$\frac{1}{4} \cdot \frac{1}{4} = \frac{1}{16}$$

and

$$\frac{1}{4} + \frac{1}{4} = \frac{2}{4} = \frac{1}{2}$$

Looks like we have our factors. As I said, look for something that when you multiply, will become the last number in the expression; and when you add or subtract, will be the middle number.

You might ask, what about 8? 8 goes into 16 twice and $2 \cdot 4 = 8$. Try working out $\frac{1}{8} + \frac{1}{8}$ and $\frac{1}{8} \cdot \frac{1}{8}$ and you will see that it is not the right answer. Sometimes it might take a while by trial and error to find the right answer, but it will be there.

Here you will end up with:

$$\left(x + \frac{1}{4}\right)\left(x + \frac{1}{4}\right)$$

which can also be written as:

$$\left(x + \frac{1}{4}\right)^2$$

When you are putting your factorization together, also think about how you want it to turn out, especially when dealing with negatives and positives. The way to tell if you have done it right is, of course, to multiply it out again. You should get the quadratic expression that you started with.

Just remember that the principle I showed you is the same with any quadratic expression that has the coefficient of 1 for x^2. It can still be done when the coefficient of x^2 is greater than 1, but it's a bit harder.

—Dr. Math, The Math Forum

Why Factor?

Dear Dr. Math,

I haven't seen this answer yet, anywhere I've looked. *Why* specifically does one have to factor out a problem? I've been explained the how, the where, and the what but not the *why*. I need a practical explanation in order to understand math or how it applies to something. Take a problem like $4b^2y^2 - 20b^2y + 24b^2 = ?$

What is the practical use of factoring this out? I know it's like unmultiplying. What applications would it serve? An example where a formula like that might be used would be helpful. Why not just solve the problem anyway? Aren't all the individual pieces needed there to solve it?

Sincerely,

Arturo

Dear Arturo,

If I understand you correctly, you want to know why the expression

$$4b^2(y^2 - 5y + 6) = ?$$

is preferable to the expression

$$4b^2y^2 - 20b^2y + 24b^2 = ?$$

Well, the main reason for identifying common factors is to let you see if there is anything recognizable in the expression that allows you to work with it further. In this case, once we get the $4b^2$ out in front, we can see that we have a standard quadratic form in y, which we can simplify even more:

$$\begin{aligned} 4b^2y^2 - 20b^2y + 24b^2 &= 4b^2(y^2 - 5y + 6) \\ &= 4b^2(y - 2)(y - 3) \end{aligned}$$

Now, why is *this* form preferable? Well, for one thing, if the "?" is a zero, as it is in standard form, then we know that

$$4b^2(y - 2)(y - 3) = 0$$

Just by looking at this, we can see that if b is nonzero, there are only two possible values of y that can make this equation true: $y = 2$ and $y = 3$.

It's also easy to see that halfway in between those two values,

$$4b^2(2.5 - 2)(2.5 - 3) = 4b^2(0.5)(-0.5)$$

the value of the expression is negative, which means that since we know that every quadratic expression is the description of a parabola, we now have enough information to draw a pretty good graph of the expression. Try doing that with the original form of the equation.

Note that this also tells us that if we vary the value of b, we can alter the shape of the parabola, but we can't move the axis of symmetry (since that would require us to change where it hits the x-axis).

In general, when mathematicians develop a technique like factoring out common terms, you can be sure it's because they've found that it turns some really tedious and error-prone task (like trying to draw the graph of a function) into a very easy one. One way to make learning math more entertaining is to try to approach the subject the way a mathematician would.

I'm not saying that you should learn to enjoy jotting down pages and pages of equations! What I mean is that you should get into the habit of looking for the easiest possible way to get from the statement of a problem to its solution, which in many cases means finding a short connection between the problem you're working on and one that you know has already been solved—whether or not you were the one who solved it.

Factoring things out is, ultimately, a technique for helping you do this.

As you've noted, you can solve problems without factoring out common terms—and you can get from New York to Los Angeles on foot if you want to, but it will take longer.

—Dr. Math, The Math Forum

Solving Quadratic Equations by Graphing

It's helpful to have the picture or graph in mind when factoring to solve a quadratic equation, but it's also quite possible to skip the factoring step and just graph the equation. Then you can read the solution of your equation from the graph.

Sometimes it helps you solve quadratic equations if you know what they look like when they're graphed. Here's an example:

You'll notice that if we factor this quadratic equation, we find two possible solutions: x is –1 or x is 3. Now look at the graph. Do you see where the parabola crosses the x-axis?

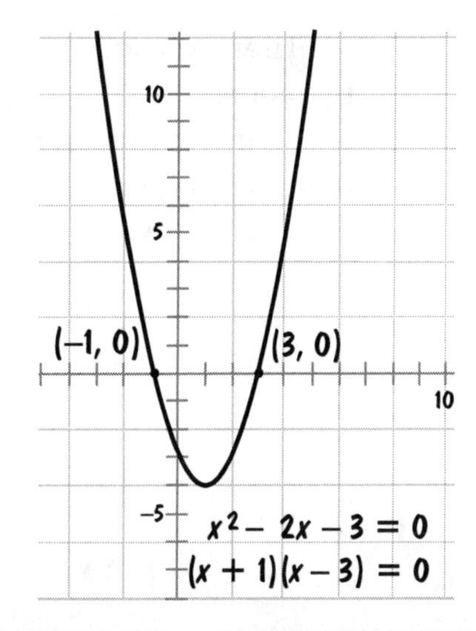

Graphing Quadratic Equations

Dear Dr. Math,

I'm doing my homework and don't understand how to do this problem, and I was hoping you could show me. How do you graph the quadratic function, $y = 3(x + 1)^2 - 20$?

Thanks for your help.

Aimee

Dear Aimee,

When doing this kind of problem, it helps to have an idea what the standard parabola $y = x^2$ looks like:

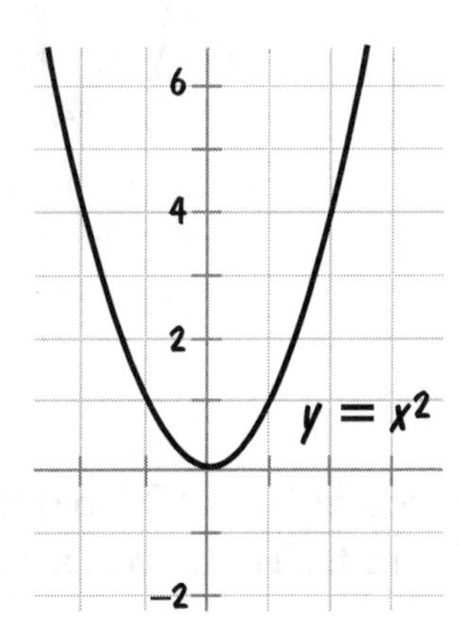

Then you see how your equation differs from the standard, piece by piece.

So in your case, let's build up your equation. If it were $y = (x + 1)^2$, it would just be the standard parabola, but moved to the left 1 unit.

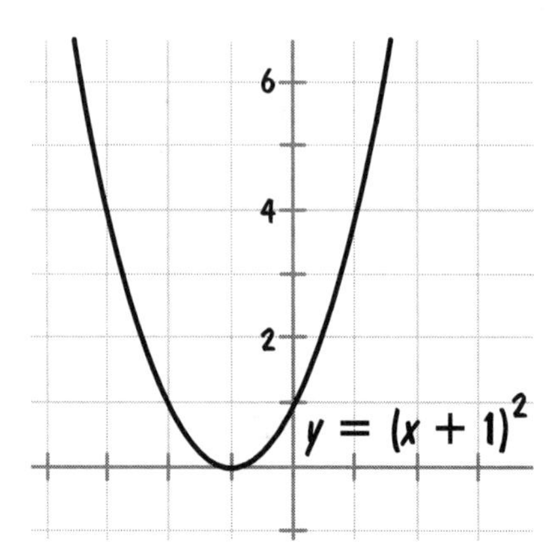

Then we multiply by 3, which takes the parabola and makes it skinnier. Notice the width of the figure above: at 2 on the y-axis, it's about 3 units wide. In the figure below, it's about 3 units wide at 6 on the y-axis, which is 3 times 2.

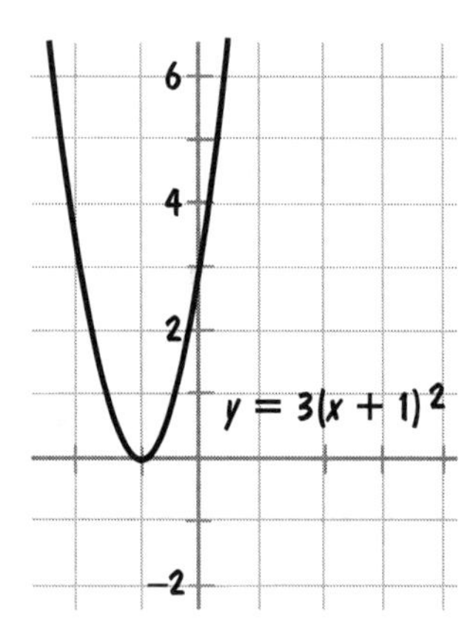

Then when you subtract 20 to get $3(x + 1)^2 - 20$, the whole picture moves down 20 units. Then you're done.

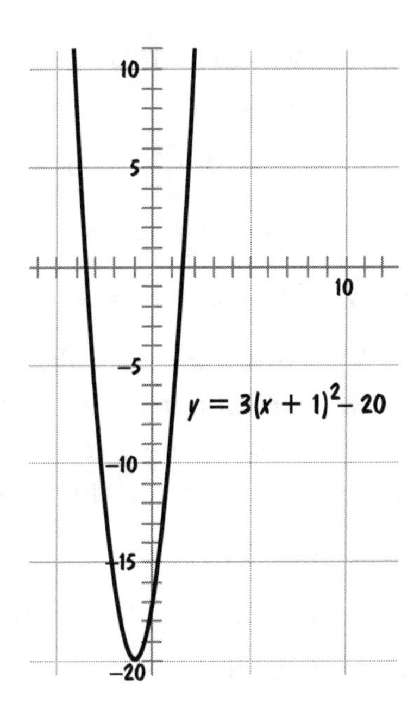

So this is a parabola centered at (–1, –20) and skinnier and lower than the standard parabola.

—Dr. Math, The Math Forum

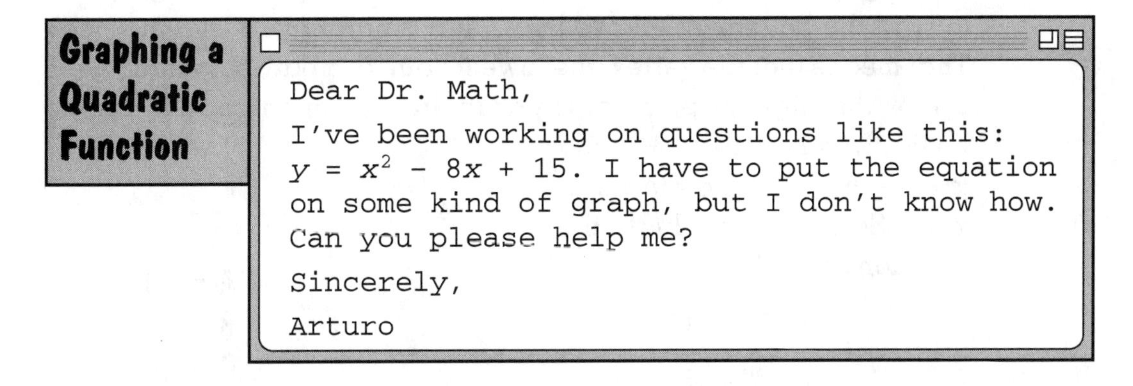

Graphing a Quadratic Function

Dear Dr. Math,

I've been working on questions like this: $y = x^2 - 8x + 15$. I have to put the equation on some kind of graph, but I don't know how. Can you please help me?

Sincerely,

Arturo

Dear Arturo,

When you want to graph a function like

$$y = x^2 - 8x + 15$$

there are two ways to do it: the hard way and the easy way.

The hard way is to start picking values for x and computing corresponding values for y:

$$x = 0, y = 0^2 - 8(0) + 15 = 15$$

$$x = 1, y = 1^2 - 8(1) + 15 = 8$$

$$x = 2, y = 2^2 - 8(2) + 15 = 3$$

and so on. The problem with this is that it can take forever, and sometimes the function wiggles in between the points you choose, so you end up drawing the wrong graph.

The easy way is to factor the equation:

$$\begin{aligned} y &= x^2 - 8x + 15 \\ &= (x - a)(x - b) \qquad \text{Where } a + b = 8, ab = 15 \\ &= (x - 3)(x - 5) \end{aligned}$$

What does this tell us? Well, for starters, it tells us that y must be zero whenever $x = 3$ or $x = 5$. (Do you see why?) So we have two points right away: (3, 0) and (5, 0).

A quadratic function like this is always shaped like a parabola. The big question is: Does it open toward the top of the graph or toward the bottom? Well, we know that a parabola is symmetrical. That means that the vertex (the lowest point of a parabola that opens up, or the highest point of a parabola that opens down) has to be halfway between $x = 3$ and $x = 5$—that is, it has to be at $x = 4$. So we can evaluate the function at $x = 4$:

$$\begin{aligned} y &= 4^2 - 8(4) + 15 \\ &= 16 - 32 + 15 \\ &= 31 - 32 \\ &= -1 \end{aligned}$$

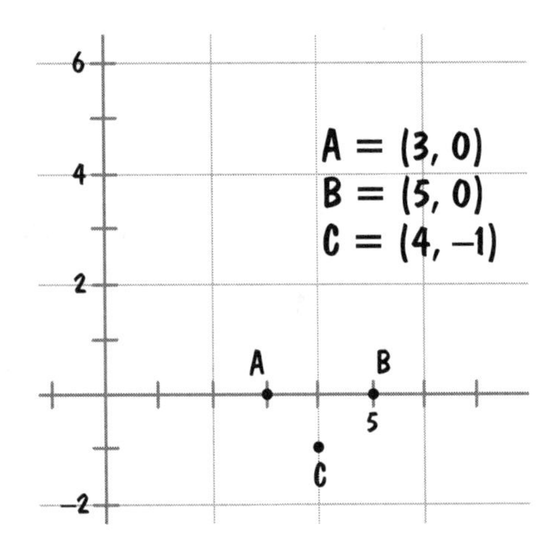

So the function has to look like this:

We can complete the graph by sketching a parabola through these three points:

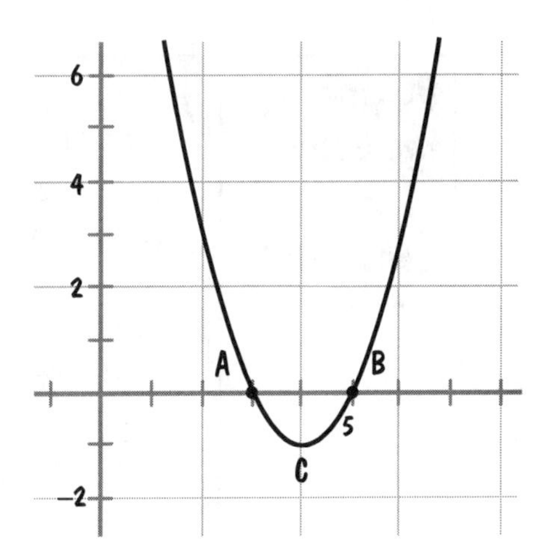

Note that the "easy" way requires you to *know* more than the "hard" way.

A lot of what you're supposed to be learning about graphs is how to guess what you're going to see before you begin plotting. In particular, the degree of a polynomial function tells you how many times it will bend:

A linear polynomial ($y = ax + b$) doesn't bend at all.

A quadratic polynomial ($y = ax^2 + bx + c$) bends one time.

A cubic polynomial ($y = ax^3 + bx^2 + cx + d$) bends twice.

And so on. Once you know how many times it has to bend, then if you can find the points at which the function crosses the x-axis, it becomes easy to draw the function. That's what I did here.

—Dr. Math, The Math Forum

Describing the Graph of an Equation

Dear Dr. Math,

Could I please have some help with this question? Thanks.

Which of the following is true of the graph of the equation $y = 2x^2 - 5x + 3$?

A. It is tangent to the x-axis.

B. It intersects the x-axis at only two distinct points.

C. It intersects the x-axis at more than two distinct points.

D. It lies completely below the x-axis.

E. It lies completely above the x-axis.

Aimee

Dear Aimee,

We can analyze this to see what we can throw out right away.

Since the coefficient of x^2 is positive, the parabola opens upward. Therefore D cannot be true.

It is only a second-degree equation, so it could not intersect the x-axis more than twice. Therefore C cannot be true.

So now we have three choices: A, B, or E.

If it's A, then when $y = 0$, there can only be one value of x that will satisfy the equation.

If it's B, then when $y = 0$, there will be two values of x that satisfy the equation.

If it's E, then when $y = 0$, there are no values of x that satisfy the equation.

To find out which it is, we will set y to zero and look for solutions.

$$2x^2 - 5x + 3 = 0$$

Factoring, we get:

$$2x^2 - 5x + 3 = 0$$
$$(x - 1)(2x - 3) = 0$$

$$x - 1 = 0 \text{ or } 2x - 3 = 0$$

$$x = 1 \text{ or } x = \frac{3}{2}$$

There are two values of x that satisfy the equation, so B is the correct answer.

—Dr. Math, The Math Forum

4 Solving Quadratic Equations by Taking Square Roots

It's always handy to have more than one way to do something. So far we have talked about solving quadratic equations by factoring and by graphing. Here is yet a third way to solve quadratic equations.

Square Roots in Equations

Dear Dr. Math,

I'm supposed to solve this: $2(x - 4)^2 - 3 = 13$.

Here's what I think:

$$2(x-4)-3=\sqrt{13} \qquad 2(x-4)-3=-\sqrt{13}$$

$$2x-8-3=\sqrt{13} \qquad 2x-8-3=-\sqrt{13}$$

$$2x-11=\sqrt{13} \quad \text{or} \quad 2x-11=-\sqrt{13}$$

$$2x=11+\sqrt{13} \qquad 2x=11+-\sqrt{13}$$

$$x=\frac{11+\sqrt{13}}{2} \qquad x=\frac{11+-\sqrt{13}}{2}$$

Did I do this right, or did I mess up at the beginning with the exponent? The exponent kind of throws me off because I'm not sure if I'm allowed to just take it out of there like that before I multiply the parentheses.

Sincerely,

Arturo

Dear Arturo,

I think you recognize that what you have done isn't right. This is good; knowing when you are wrong is the first step toward learning to get it right. So let's take another look at that equation:

$$2(x - 4)^2 - 3 = 13$$

Every equation has certain difficulties that it puts in the way of finding the answer. The biggest difficulty that this equation gives us is that exponent. It's natural that you should want to get rid of the exponent as quickly as possible, but to do that you have to take the square root, and now there's a barrier in the way of just taking the square root, because whatever you do in an equation you have to do to *all* of both sides. If we take the square root as our first step, the result is:

$$\sqrt{2(x - 4)^2 - 3} = \sqrt{13}$$

As you can see, this isn't very helpful because all that stuff that you're supposed to take the square root of on the left-hand side is pretty messy. That means that we need to do a little more rearranging of the equation before we can take any square roots.

Now you can easily take the square root of $(x - 4)^2$ or even of $2(x - 4)^2$, but that's not what you have on the left side of the equation: you have that -3 term also. So before you can go taking square roots, you have to get rid of that -3. You can do that by adding 3 to both sides of the equation. This gives you:

$$2(x - 4)^2 = 16$$

Next you probably want to get rid of the factor of 2 on the left side. That's easy since you can just divide both sides by 2. This gives:

$$(x - 4)^2 = 8$$

Now you have something squared all by itself on the left side, and you can go ahead and take square roots. So you get:

$$(x - 4) = \sqrt{8} \text{ or } (x - 4) = -\sqrt{8}$$

This is because there are *two* numbers whose square is 8, one of them positive and the other negative (which you seem to already know about).

This gives the two solutions for x:

$$x = 4 + \sqrt{8} \text{ and } x = 4 - \sqrt{8}$$

You may want to (or your teacher may want you to) simplify the $\sqrt{8}$ further—it can be written as $2\sqrt{2}$. (Do you see why?) This makes the answers

$$x = 4 + 2\sqrt{2} \text{ and } x = 4 - 2\sqrt{2}$$

—Dr. Math, The Math Forum

5 Solving Quadratic Equations by Completing the Square

Completing the square is just one of many tools for dealing with quadratic equations. As with any set of tools, in general, you try to use whichever one will require the least effort.

For example, there's nothing easier than simply recognizing a common pattern. We can try factoring, which requires guess-and-check. But maybe our equation doesn't match any pattern that we recognize, and we can't find an easy factorization.

But if the lead coefficient is a square, a^2; and if the linear coefficient is a multiple of $2a$; then we should be able to complete the square without much trouble.

So why complete the square? Because sometimes it's the least awkward tool that will get the job done.

Completing the Square

Dear Dr. Math,

$2x^2 = x + 5$. How do I complete the square?

Aimee

Dear Aimee,

Let's start with an example:

$$3x^2 = 6x + 12$$

First, to solve a quadratic equation in one variable like this one by completing the square, get just the quadratic term and the linear term (if there is one) on one side of the equation. Here, subtract 6x from both sides to get:

$$3x^2 - 6x = 12$$

Second, divide each term in the equation by the coefficient of the squared term so its new coefficient is 1.

$$x^2 - 2x = 4$$

Third, divide the coefficient of the linear term (the x term) by 2 and square the result.

$$\left(\frac{-2}{2}\right)^2 = 1$$

Then add this number to both sides. Now you have:

$$x^2 - 2x + 1 = 4 + 1$$

Fourth, simplify the right side:

$$x^2 - 2x + 1 = 5$$

The left side is now a perfect square trinomial. Therefore, you can factor it into the binomial squared, $(x - 1)^2$, and write the equation as follows:

$$(x - 1)^2 = 5$$

Fifth, take the square root of both sides of the equation, noting that the square root of a number can be positive or negative.

$$(x - 1) = \pm\sqrt{5}$$

Sixth (and last), separate and solve each linear equation.

$$(x - 1) = \sqrt{5} \quad \text{or} \quad (x - 1) = -\sqrt{5}$$

$$x = \sqrt{5} + 1 \qquad\qquad x = -\sqrt{5} + 1$$

I'll use the same steps in this problem. See if you can follow each step:

$$5x^2 - 1x - 10 = 0$$

$$5x^2 - 1x = 10$$

$$x^2 - \frac{1}{5}x = 2$$

Here's the tricky part: Divide $-\frac{1}{5}$ by 2 and square the result. Add the number to both sides.

$$x^2 - \frac{1}{5}x + \frac{1}{100} = 2 + \frac{1}{100}$$

$$\left(x - \frac{1}{10}\right)^2 = \frac{201}{100}$$

$$\left(x - \frac{1}{10}\right) = \pm\sqrt{\frac{201}{100}}$$

$$= \pm\frac{\sqrt{201}}{10}$$

$$x - \frac{1}{10} = \frac{\sqrt{201}}{10} \quad \text{or} \quad x - \frac{1}{10} = -\frac{\sqrt{201}}{10}$$

$$x = \frac{\sqrt{201}}{10} + \frac{1}{10} \qquad\qquad x = -\frac{\sqrt{201}}{10} + \frac{1}{10}$$

$$x = \frac{\sqrt{201} + 1}{10} \qquad\qquad x = \frac{-\sqrt{201} + 1}{10}$$

This solution is not exactly pretty, but it's correct.

—*Dr. Math, The Math Forum*

Factoring without Guess-and-Check

Dear Dr. Math,

I am now learning how to solve quadratic equations. I found out how, but I am still wondering how to do one thing. It is part of the factoring. I will use the example $x^2 + 5x + 6 = 0$.

My mom taught me that you find two factors of the 6 that add up to the 5 in $5x$. In this case it can be 2 and 3. She told me to then make it into $(x + 3)(x + 2) = 0$.

I am wondering if there is any way to find two factors of 6 (or whatever number it is) that add up to 5 (or the other number) without using guess-and-check. If there is no way to do that, then how do you factor it when the two factors are not whole numbers? I hope you can help me with this.

Sincerely,

Arturo

Dear Arturo,

Interestingly, guess-and-check is the most straightforward way to proceed if you have reason to believe that the roots are integers, and what makes it reasonable is that there usually aren't all that many possible guesses! This is because for any given integer, there are a limited number of ways to multiply two other integers to get it.

For example, consider something like:

$$x^2 + 13x - 68$$

If you break 68 into prime factors,

$$68 = 2 \cdot 2 \cdot 17$$

it becomes clear that the *only* possible ways to factor –68 are:

$$-1 \cdot 68$$

$$1 \cdot -68$$

$-2 \cdot 34$

$2 \cdot -34$

$-4 \cdot 17$

$4 \cdot -17$

and only one of these pairs of factors can add up to 13. So it's guess-and-check, but when done correctly, it's a very *educated* kind of guess-and-check.

But there is a way of avoiding guess-and-check, called "completing the square," which makes use of two very common patterns:

1. $(x + a)^2 = x^2 + 2ax + a^2$
2. $(x + a)(x - a) = x^2 - a^2$

You're going to be seeing a lot of these as you continue on in math, so you may as well take this opportunity to become friends with them.

How do we use them? First, we complete the square by adding the square of half the coefficient of x to both sides of the equation. Using your example equation,

$$x^2 + 5x + 6 = 0$$

$$x^2 + 5x + \left(\frac{5}{2}\right)^2 + 6 = \left(\frac{5}{2}\right)^2$$

$$\left(x + \frac{5}{2}\right)^2 + 6 = \frac{25}{4}$$

Now we subtract it from both sides to get

$$\left(x + \frac{5}{2}\right)^2 + 6 - \frac{25}{4} = 0$$

$$\left(x + \frac{5}{2}\right)^2 + \frac{24}{4} - \frac{25}{4} = 0$$

$$\left(x + \frac{5}{2}\right)^2 - \frac{1}{4} = 0$$

And since $\frac{1}{4}$ is a square (of $\frac{1}{2}$), we can use our second pattern:

$$\left(x+\frac{5}{2}\right)^2-\frac{1}{4}=0$$

$$\left(x+\frac{5}{2}+\frac{1}{2}\right)\left(x+\frac{5}{2}-\frac{1}{2}\right)=0$$

$$(x+3)(x+2)=0$$

It has a certain elegance, doesn't it? But as you can see, the numbers can get a little hairy, which is why people tend to prefer the guess-and-check method. It offers fewer chances for making silly mistakes.

But the quickest way to factor a quadratic equation is to use the quadratic formula. Again, using your example,

$$x^2 + 5x + 6 = 0$$

the quadratic formula says that the roots are:

$$x=\frac{-b\pm\sqrt{b^2-4ac}}{2a}$$ Where $ax^2 + bx + c = 0$. Here $a = 1$, $b = 5$, and $c = 6$.

$$\begin{aligned}x&=\frac{-5\pm\sqrt{5^2-4\cdot1\cdot6}}{2\cdot1}\\&=\frac{-5\pm\sqrt{25-24}}{2}\\&=\frac{-5\pm1}{2}\\&=\frac{-4}{2},\frac{-6}{2}\\&=-2,-3\end{aligned}$$

This tells us that the original equation is zero when x = –2 or when x = –3; so

$$x^2 + 5x + 6 = (x + 2)(x + 3)$$

Note that you have to be careful with signs when using this method! It's easy to get them mixed up.

The quadratic formula is by far the fastest way to factor a quadratic equation, and it has the advantage that it works even when the roots aren't integers.

—Dr. Math, The Math Forum

COMPLETING THE SQUARE AND THE QUADRATIC FORMULA

Completing the square is the method by which one usually derives the quadratic formula. That is, if $ax^2 + bx + c = 0$, and a, b, and c are real numbers, then first we get rid of the a by dividing through:

$$x^2 + \frac{b}{a}x + \frac{c}{a} = 0$$

Now we complete the square. First, subtract $\frac{c}{a}$ from both sides:

$$x^2 + \frac{b}{a}x = -\frac{c}{a}$$

Divide the coefficient of x (that's $\frac{b}{a}$) by 2 (to get $\frac{b}{2a}$) and square it (to get $(\frac{b}{2a})^2$), then add it to both sides of the equation:

$$x^2 + \frac{b}{a}x + \left(\frac{b}{2a}\right)^2 = \left(\frac{b}{2a}\right)^2 - \frac{c}{a}$$

Find a common denominator and simplify the right side of the equation:

$$x^2 + \frac{b}{a}x + \left(\frac{b}{2a}\right)^2 = \left(\frac{b}{2a}\right)^2 - \frac{c}{a}$$

$$= \frac{b^2}{4a^2} - \left(\frac{c}{a} \cdot \frac{4a}{4a}\right)$$

$$= \frac{b^2}{4a^2} - \frac{4ac}{4a^2}$$

$$= \frac{b^2 - 4ac}{4a^2}$$

We have a perfect square on the left, so taking the square root of both sides, we get:

$$x + \frac{b}{2a} = \frac{\pm\sqrt{b^2 - 4ac}}{\sqrt{4a^2}} = \frac{\pm\sqrt{b^2 - 4ac}}{2a}$$

Subtracting $\frac{b}{2a}$ from both sides, we get the familiar

$$x = \frac{-b \pm \sqrt{b^2 - 4ac}}{2a}$$

• •

Geometrically Completing the Square

Dear Dr. Math,

I've been searching everywhere to find the steps for geometrically completing the square. Do you know them?

Aimee

Dear Aimee,

Start with a rectangle. Our job is to find a square with equal area. So that we have the same picture in our heads, draw a rectangle on a piece of paper. Make it 3 inches by 2 inches, with the 3 inches horizontal.

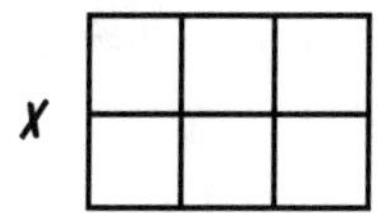

Mark the vertical side, on the left, x. Mark off x on the horizontal side, starting on the left. Now you have a square with sides x.

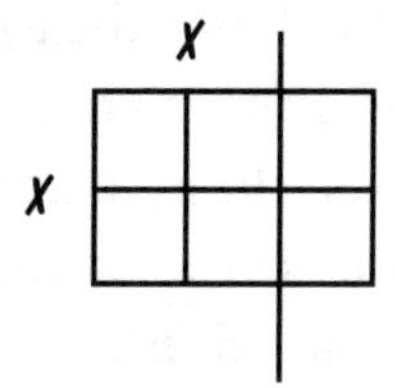

Draw the vertical line to make the square. On the top, put 7 on the remaining, shorter, segment. (It could be a number other than 7, but this is my example. Bear with me!) The small rectangle has area 7x. Right?

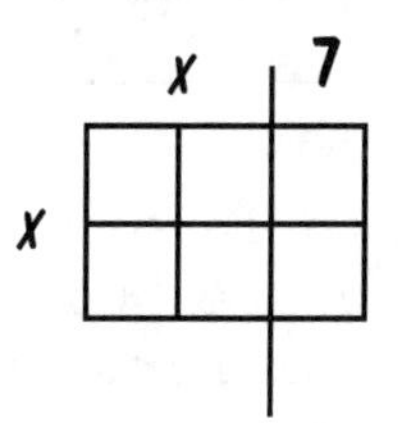

Okay, we have a rectangle whose area is $x^2 + 7x$. To complete the square algebraically, we write:

$$x^2 + 7x + \left(\frac{7}{2}\right)^2 - \left(\frac{7}{2}\right)^2 = \left(x + \frac{7}{2}\right)^2 - \left(\frac{7}{2}\right)^2$$

Geometrically, divide the thin rectangle on the right by a vertical line down the middle. Move the rightmost half down and to the left. If you rotate it 90 degrees, it will just fit under the square.

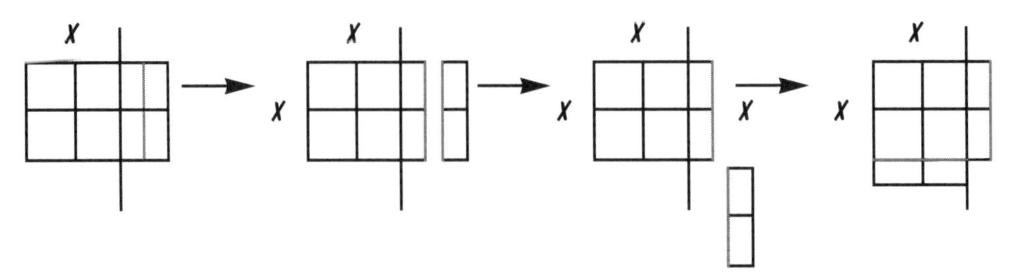

Now you see a figure with an area equal to the original rectangle but arranged so that we can complete the square. What's missing is the little square down in the lower right-hand corner. What is its area? Well, you will see that its area is $(\frac{7}{2})^2$. So, returning to the original question, to find a square with area equal to the original rectangle, all we must do is to start with two squares and construct their difference. This can be done by a geometrical procedure, too. The two squares have sides $x + \frac{7}{2}$ and $(\frac{7}{2})^2$.

Neat, huh? All of this was known to the Vedic Indians (Asian Indians), quite a long time before Euclid. It's in their religious book, the *Sulvasutra*.

—Dr. Math, The Math Forum

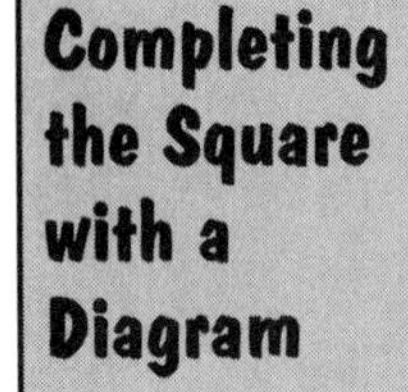

Dear Dr. Math,

My question involves completing the square, but instead of the usual form of brackets and formulae, I have to do it something like this:

x^2 | $3x$ | $3x$

I don't know how to draw a diagram for completing the square for: $x^2 - 6x + ?$

The answer is $x^2 - 6x + 9$, but I can't draw the diagram. My teacher told me the only hint we get is that it overlaps. I can't seem to figure it out. Can you help?

Sincerely,

Arturo

Dear Arturo,

Try this diagram:

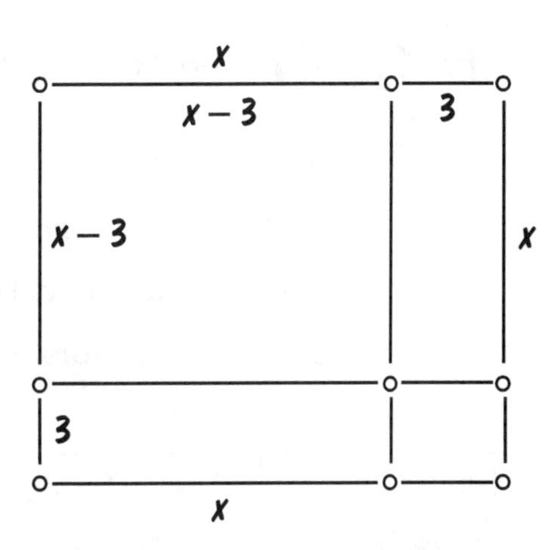

The lower right has area 3^2. The entire lower edge rectangle (small rectangle and square together making a longer rectangle) has area $3 \cdot x$, and likewise the entire right edge rectangle has area $3 \cdot x$. They overlap in the 3-by-3 square in the lower right corner.

—***Dr. Math, The Math Forum***

6 The Quadratic Formula

The quadratic formula: $x = \dfrac{-b \pm \sqrt{b^2 - 4ac}}{2a}$ can be used where the equation to be solved is $ax^2 + bx + c = 0$. Some people are comfortable with it and use it all the time even though other methods would work easily. Others prefer to use it only as a last resort.

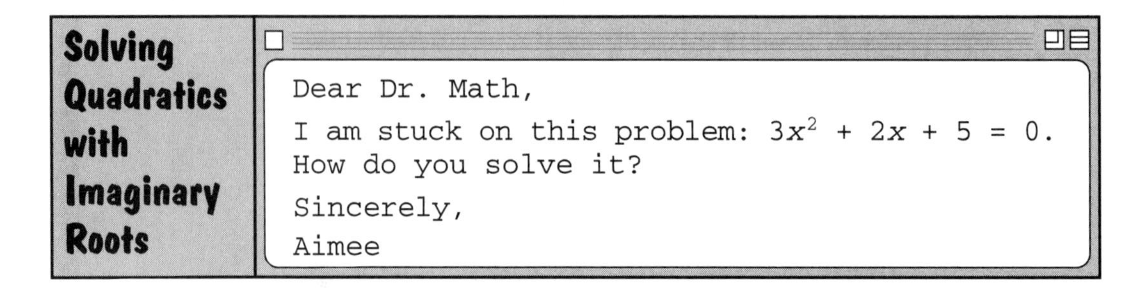

Solving Quadratics with Imaginary Roots

Dear Dr. Math,

I am stuck on this problem: $3x^2 + 2x + 5 = 0$. How do you solve it?

Sincerely,

Aimee

Dear Aimee,

There's a good reason for you to be stuck on this problem, because it doesn't have a solution—unless you are learning about imaginary numbers! I'll give you several ways to look at this problem.

Your problem looks like this:

$$3x^2 + 2x + 5 = 0$$

You may be trying to solve it by factoring. If you are, you've probably been trying everything you can think of, and nothing works. That's the problem with factoring: You can tell when you've done it, but you can't easily tell when it can't be done—and most quadratic equations I could make up can't be factored. In real life we usually solve first, then factor if we need to by using the solution to give us the factors. The quickest way is to use the quadratic formula:

$$x = \frac{-b \pm \sqrt{b^2 - 4ac}}{2a}$$

where the equation to be solved is $ax^2 + bx + c = 0$.

For this problem, $a = 3$, $b = 2$, and $c = 5$, so the answer is:

$$x = \frac{-2 \pm \sqrt{4-60}}{6}$$

Since $4 - 60 = -56$, you can't take the square root (as a real number), so we know there is no solution. That saves us a lot of work factoring.

(If you know about imaginary numbers, you can simplify $\sqrt{-56}$ to $2 \cdot i \cdot \sqrt{14}$ and get a pair of complex solutions. For now, I'll assume we only care about real solutions.)

Suppose you were trying instead to solve the equation by completing the square. Then you would have written something like this:

$$3x^2 + 2x + 5 = 0$$

$$3\left(x^2 + \frac{2}{3}x\right) + 5 = 0$$ Factor 3 out of the x terms.

$$3\left(x^2 + \frac{2}{3}x + \frac{1}{9}\right) + 5 - \frac{1}{3} = 0$$ Add and subtract $\frac{1}{9}$ in the parenthesis to make a square; note that it becomes $\frac{1}{3}$ outside the parenthesis, because the parenthesis is multiplied by 3.

$$3\left(x + \frac{1}{3}\right)^2 + \frac{14}{3} = 0$$ Write it as a square.

This tells you that there is no solution because $(x + \frac{1}{3})^2$ is never less than zero, and it would have to be $-\frac{14}{9}$ to make the left side zero. Do you see why this is?

What does it mean for this equation not to have a solution? Try graphing $y = 3x^2 + 2x + 5y$ and you'll see that it's a parabola that never crosses the x-axis; that is, there is no x for which it equals zero. In fact, if you look back at the solution by completing the square, the bottom of the parabola (its lowest value) is at $x = -\frac{1}{3}$, when $x + \frac{1}{3}$ is zero, and at that point $y = \frac{14}{3}$. So even though completing the square didn't give you a solution, it gave us lots of information about the equation.

One more thought: It could be that you couldn't solve this because you copied the equation wrong! Make sure you check, because a change of sign could change everything.

I may have gone through all this too fast for you, because I'm not sure which parts to concentrate on. The main idea is to see how you can tell when a quadratic equation can't be solved. Remember, if you can't solve something because it can't be solved, you haven't failed! If you can show that it has no solution, that's the answer!

—Dr. Math, The Math Forum

The Discriminant of Quadratic Equations

Dear Dr. Math,

Why do some people use a triangle to represent $b^2 - 4ac$ in the quadratic formula?

Arturo

Dear Arturo,

This is all about how to solve quadratic equations, such as:

$$40x^2 + 5x + 63 = 0$$

People learned how to solve lots and lots of equations like this, and then, a few hundred years ago, somebody asked what do the 40, the 5, and the 63 have to do with it? Can we find rules about what the answer is for every possible combination of the three coefficients? So one day, they set out to solve the equation:

$$ax^2 + bx + c = 0$$

But the answers they got depended (obviously) on a, b, and c, the coefficients. There were several different ways the problem could turn out; and what is most interesting, exactly which way it went was decided by the formula $b^2 - 4ac$. People decided to call that expression the "discriminant," because it discriminated between the various solutions. The Greek letter delta was chosen as its symbol: Δ. Delta looks like a triangle, but it is a Greek letter, not a geometric shape.

The expression $ax^2 + bx + c$ is an interesting one. With some careful math, it can be split into two parts:

$$a \cdot \text{[varying part that's never negative]} - \text{[fixed part]}$$

The fixed part is $\frac{\Delta}{4a}$, so you see the connection!

If delta is negative, the whole combination has absolutely no chance of being zero, so the equation $ax^2 + bx + c = 0$ cannot be solved. To see this, put the whole thing over a common denominator, and you get:

$$\frac{4a^2 \cdot [\text{varying part, never negative}] - \Delta}{4a}$$

If delta is negative, –delta will be positive, and $4a^2$ is certainly positive, so the numerator is going to be at least as big as –delta, so it can never be zero, regardless of the denominator.

If delta is positive, there will be two different values for x at which the whole expression becomes zero. This is the so-called generic case. To see this, I have to tell you that the "varying part" is just $(x + \frac{b}{2a})^2$, which takes any desired positive value exactly twice.

If delta is zero, the only value of x that will make the expression zero is $x = -\frac{b}{2a}$, so you have one solution.

A much nicer way of understanding the solution is to think of it in terms of graphs. You must know what the graph of the expression

$$ax^2 + bx + c$$

looks like, in which case the three cases become obvious. Look in your textbook. There should be diagrams showing you the situation. There will be a U-shaped curve, called a parabola. There will be a horizontal line that's the x-axis. Each point on the x-axis stands for a possible x value. The corresponding value of $ax^2 + bx + c$ is just how high the point on the parabola is from the x-axis. When the curve actually crosses the x-axis, the height is zero, of course, so the value is zero. The question then becomes does this graph cross the x-axis? At how many points?

The a, b, and c determine the position of the parabola. The "a" determines whether the parabola is arch-shaped (a is negative) or U-shaped (a is positive). Delta determines whether the parabola crosses the x-axis (delta is positive), just touches the x-axis (delta is zero), or misses the x-axis completely (delta is negative).

—Dr. Math, The Math Forum

Resources on the Web

Learn more about quadratic equations at these Math Forum sites:

Algebra Problem of the Week: The Christmas Gifts

mathforum.org/algpow/solutions/solution.ehtml?puzzle=58

A grandfather doles out his annual Christmas gifts to his beloved grandchildren in a most unusual way—mathematical, of course.

Algebra Problem of the Week: The Continued Fraction

mathforum.org/algpow/solutions/solution.ehtml?puzzle=26

Find the value of a given continued fraction.

Discrete Math Problem of the Week: The Fourth of July Parade

mathforum.org/dmpow/solutions/solution.ehtml?puzzle=1

After the parade, the people on the float I was on shook hands with one another. The mayor came over and shook hands with only the people he knew. How many people did he know if there were 1,625 handshakes altogether?

Algebra Problem of the Week: The Length of Larry's Rectangle

mathforum.org/algpow/solutions/solution.ehtml?puzzle=59

Larry wants to know the length of his rectangle. Can you help him?

Algebra Problem of the Week: Valentine Candy

mathforum.org/algpow/solutions/solution.ehtml?puzzle=24

A vendor who sells Valentine candy boxes wants to raise the price. For every twenty-five-cent increase he loses two customers. At what price should he sell the boxes to achieve the maximum profit?

Glossary

binomial A polynomial with exactly two terms.

coefficient In an algebraic term, a number that's multiplying the variable. In the term $4x$, 4 is the coefficient; in $-9x^3$, the coefficient is -9; in x, the coefficient is 1. Coefficients are not constants, because they're tied to the variable and vary with it.

coincident Occurring at the same time and place. If two lines are coincident, they lie on top of each other—meaning they're really the same line.

common denominator A quantity into which all the denominators of a set of fractions may be divided without a remainder. The least common denominator is when this quantity is the smallest such quantity possible.

common factor When comparing two or more numbers or expressions, a factor that appears in each.

consistent A system of equations is consistent if there is at least one solution that satisfies all equations in the set. If the equations are linear, the lines cross at one point.

constant Something that is unchanging or *invariable,* that is, not a variable. For example, in $2x + 3$, x is a variable, 2 is a coefficient attached to the variable, and 3 is a constant.

degree In a polynomial, the largest sum of the exponents of any term is the polynomial's degree. Quadratic equations have degree 2; the expression $4x^3y^2 + 5x^2y - 4x + 6$ has degree 5.

dependent A system of equations is dependent if there are many different solutions for it. If the equations are linear, they all describe the same line.

dependent variable The output variable of a function. Once you pick the value of the input variable, that determines the value of the output—the output is *dependent* on the input variable.

difference of squares Refers to the pattern $a^2 - b^2 = (a + b)(a - b)$.

elimination To eliminate is to get rid of or remove. In systems of equations, elimination is the process of combining equations so that a variable drops out, to make it easier to solve.

equation An equation is a statement that two things are equal; a mathematical sentence.

evaluate To work out an expression using given values for any variables. This usually results in a single, numeric answer.

expression A symbol, number, or combination of either or both, representing a quantity or relation between quantities. If an equation is a mathematical sentence, an expression is a mathematical phrase.

factor The numbers multiplied together to get another number are its factors. For example, $4 \cdot 3 = 12$, so 3 and 4 are factors of 12. However, they're not its only factors. Other factors of 12 are 1, 2, 6, and 12. (Another way of defining a factor is a number that divides evenly into the number you're factoring.)

formula A common principle or relation expressed in algebraic terms, usually as an equation or function.

function A function is a relation between two or more variables, such that for any value of the independent variable(s), there is exactly one value for the dependent variable.

inconsistent If a set of equations is inconsistent, there is no solution for the set. If the equations are linear, they describe parallel lines.

independent variable The input variable of a function. This can be thought of as *independent* because you can freely and independently change it to whatever you want.

intercept A line's x-intercept is the place where it crosses the x-axis, and its y-intercept is the place where it crosses the y-axis. You might think of the line catching an axis, as if it were intercepting a ball.

intersection To intersect is to cross or overlap. In systems of two equations in two unknowns, intersection is the process of solving both equations for one variable and setting the resulting expressions equal to each other.

linear equation A linear equation is an equation with two variables that describes a line. Technically, it's called an equation of degree 1, because each variable has an exponent of 1 (which doesn't usually need to be written out: that is, x^1 is the same thing as x).

monomial An algebraic expression consisting of only one term.

overdetermined See **inconsistent.**

parabola A plane curve resulting from the intersection of a right cone and a plane parallel to the side of the cone. A parabola can also be described as the set of points equidistant from a line, called the directrix, and a point not on the line, called the focus.

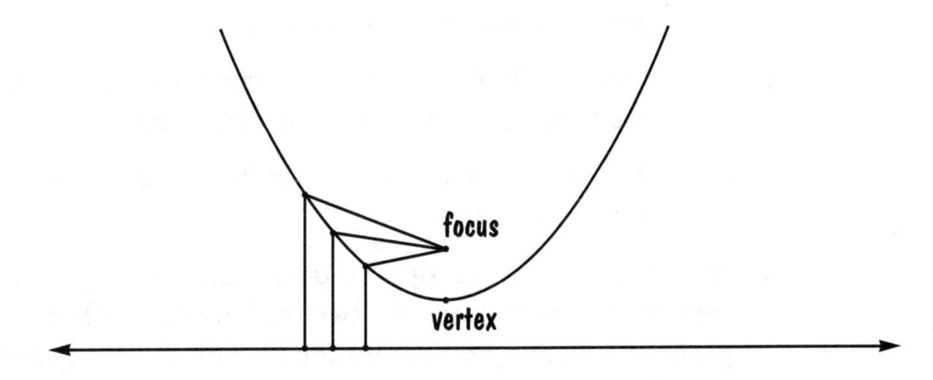

polynomial An algebraic expression consisting of one or more terms added (or subtracted) together, each term consisting of a constant multiplier and one or more variables raised to nonnegative integral powers.

quadratic Polynomial of degree 2.

quadratic equation A quadratic *equation* is an equation you get when you set a quadratic *function* equal to zero or some other constant.

quadratic formula The formula for the roots of a quadratic equation:

$$x = \frac{-b \pm \sqrt{b^2 - 4ac}}{2a}$$

quadratic function A quadratic function is a polynomial function of degree 2.

scale To multiply an equation by a fixed amount, making its components larger or smaller but keeping the same relations between quantities.

simplify To make simple. In an algebraic expression, to work out as fully as possible. Compare **evaluate.**

simultaneous equations If you do two things simultaneously, you are doing them at the same time. Simultaneous equations are equations that are true at the same time.

slope Slope is a number that shows how steeply a line slants. We often use the letter *m* to stand for a line's slope.

square The product of a number and itself; 9 is the square of 3, and $4x^2$ is the square of 2x.

substitution The act of replacing one thing with another. In a system of two equations in two unknowns, substitution is the process of solving one equation for one variable in terms of the other, and substituting the result into the second equation to get one equation in one variable.

system of equations See **simultaneous equations.**

trinomial A polynomial with exactly three terms.

undefined slope A slope with zero in the denominator; that is, vertical. Since lines with such slopes never go to the right, the "run" of "rise over run" is zero, which makes the slope undefined.

underdetermined See **dependent.**

unknown The unknown in a problem is the value for which you're solving, usually represented with a variable.

variable A symbol, usually a lower-case letter, representing a value you don't yet know.

vertex On a parabola, the point directly between the focus and the directrix. The axis of the parabola runs through this point, perpendicular to the directrix.

x-intercept A line's x-intercept is the place where it crosses the x-axis; that is, the value of x when $y = 0$.

y-intercept A line's y-intercept is the place where it crosses the y-axis; that is, the value of y when $x = 0$.

Index

F

G

H

I

Q

V

W

X

Y

Z